Also by Jay Ingram

The Science of Why
Volume 4

Answers to Questions About Science Facts, Fables, and Phenomena

Jay Ingram

PUBLISHED BY SIMON & SCHUSTER
NEW YORK   LONDON   TORONTO   SYDNEY   NEW DELHI

SIMON &
SCHUSTER
CANADA

Simon & Schuster Canada
A Division of Simon & Schuster, Inc.
166 King Street East, Suite 300
Toronto, Ontario M5A 1J3

This Simon & Schuster Canada edition November 2019

SIMON & SCHUSTER CANADA and colophon are trademarks
of Simon & Schuster, Inc.

For information about special discounts for bulk purchases,
please contact Simon & Schuster Special Sales at 1-800-268-3216
or CustomerService@simonandschuster.ca.

Illustrations by Tony Hanyk (tonyhanyk.com) and Elizabeth Whitehead

Manufactured in the United States of America

10   9   8   7   6   5   4   3   2   1

Library and Archives Canada Cataloguing in Publication
    Title: The science of why. Volume 4 : answers to questions about science facts, fables,
and phenomena / by Jay Ingram.
    Names: Ingram, Jay, author.
    Identifiers: Canadiana (print) 2019009690X | Canadiana (ebook) 20190096926 |
    ISBN 9781982130893 (hardcover) | ISBN 9781982130909 (ebook)
    Subjects: LCSH: Science—Popular works. | LCSH: Science—Miscellanea.
Classification: LCC Q162 .I554 2019 | DDC 500—dc23

ISBN 978-1-9821-3089-3
ISBN 978-1-9821-3090-9 (ebook)

To the Men of the Road, who keep things light

Contents

Part 3: Peculiar Phenomena

Part 4: Curiosities and Oddities

The Science of Why

Volume 4

Part 1
Bodily Puzzles

Why do we itch?

WE ALL KNOW HOW AN ITCH FEELS, but there's still a lot we don't know about how it works. An observation made 2,000 years ago by the Buddhist philosopher Nāgārjuna suggests the complexity of the sensation: "There is pleasure when an itch is scratched. But to be without an itch is more pleasurable still."

I'm itching for you to invite me over.

There are many serious diseases in which itch is constant, extremely unpleasant, and very difficult to treat. Fortunately, most of us only experience the transitory itch from an insect bite or a wool sweater, but even those relatively trivial itches are tough to describe.

For example, the feeling of pain and itch are similar. They're both irritating and you want them to go away, but each has a unique quality.

Did You Know . . . That famous portrait of Napoleon Bonaparte with one hand in his waistcoat and the other behind his back? Some scientists suspect that his hands are hidden because he's scratching himself. That's the problem with standing there forever as the painter paints: you just get itchy. But many believe Napoleon did have some sort of irritating skin condition. For a long time scabies, a disease caused by tiny mites, was thought to be the cause, but most medical experts doubt that now, and the cause of Napoleon's itch is still unknown.

For some time scientists believed that pain and itch were dependent on the same nerve pathways, pain simply being an amplified version of itch. The fact that scratching actually causes pain—enough pain to subdue the itch at least temporarily—can be interpreted as amping up the stimulation enough to cross the threshold between itch and pain. Some compounds that relieve pain, like opioids, can at the same time cause itchiness, as if the threshold were crossed in the other direction. And those rare and unfortunate individuals who are insensitive to pain, a life-threatening condition, also do not experience itch. However, the belief that pain and itch share the same neural infrastructure has been largely abandoned. The itch network has its own specific nerves and transmitter molecules, even though there is some interplay between those and their counterparts for pain.

Of course the two are very different. Itch is confined to the skin, whereas pain can occur virtually anywhere in the body. And our response to pain is withdrawal—snatching your hand away from the hot stove—but we react to an itch by attacking it, usually with our fingernails. Both reactions make sense, especially if you think that part of the reason we scratch is to remove anything, like a biting insect, that is responsible for the itch.

Interestingly, if we don't scratch hard enough to cause pain, scratching an itch can, as Nāgārjuna claimed, bring pleasure.

It makes sense to think of an itch as a local event on the skin, but that's not true. If you were to track the itchiness of, say, a mosquito bite, it starts at the site of the bite, where the body recognizes the chemicals released by the mosquito as foreign and reacts by releasing histamine, a notorious itch promoter. Histamine causes local nerves to fire, and those signals travel to the spinal cord, then the brain, where the impulses of those nerves are registered as an itch. So that itch on your arm is really in your brain, a very hard place to scratch.

(I think the idea that itch is centered in the brain explains something I've noticed: if I've been out in the woods and have collected several mosquito bites, even if most of them are quiescent, scratching the one that's itchy makes *all* of them start to itch. Again, the explanation lies in the brain, not in the individual bites.)

Did You Know . . . If you are itchy, there are several things you can do. Over-the-counter anti-itch remedies can help, but there are a couple of remedies that come from the scientific literature, too. For instance, seventy years ago scientists noted that light pinpricks around the site of an itch eliminated it for up to forty-five seconds. This couldn't have been a substitution of pain for itch because the mild pain of the pinprick disappeared thirty seconds before the itch returned. Others have noticed that pressing on the skin around the itch site, rather than direct scratching, also dampens the itch sensation.

More evidence that the itch is in your mind comes from experiments showing that itch can be contagious (although apparently pain—also in your brain—isn't contagious: just another difference between the two). When volunteers were shown videos of other people scratching or images of insects on their skin, they scratched themselves much more than they did when watching a neutral video. Interestingly they didn't necessarily scratch the same place on their bodies as the individuals in the video did; most of the time they just scratched their heads. But it was a real effect and raises the question of why this should happen. It resembles contagious yawning, where even the word "yawn" can make people do it. There is apparently no evidence—yet—that just reading about itchiness can make you itchy. Is there?

Science is an itch I just have to scratch.

 TRY THIS AT HOME! A scientist named Theodore Cornbleet studied a group of itchy volunteers and found that whenever they scratched, the length of the scratch varied depending on the location. Itchiness on the tips of the fingers provoked very short scratches, roughly 2 millimeters (about one-sixteenth of an inch) long, but itchy spots on the back were attacked with megascratches stretching 80 millimeters (more than three inches). Cornbleet argued that this happened because touch sensitivity varies all over the body, and where the sensitivity is less, as on the back, longer scratches are needed to influence enough neural receptors to affect the itch.

You can test your own touch sensitivity: Bend a paper clip into a U shape, then test (without looking) how close together the two ends can be before you can't tell them apart. Do that for your fingertips as well as your back, your calf, the bottom of your foot. Each will be different.

What does earwax tell us about ourselves . . . and blue whales?

EARWAX RARELY MAKES IT INTO DAILY CONVERSATION. But it deserves to, not just because we all have it, but because in some animals it's an amazing record of that animal's life.

Earwax makes people cringe, but why? Vomit and feces disgust us for a good reason: they're full of hostile bacteria and viruses. But earwax, despite our revulsion, is exactly the opposite: it's part of a natural cleansing process. And it is essential to healthy ears.

As flakes of dead skin slough off the inside of the ear, they mix with secretions from the cells lining the canal: waxes and oils containing antimicrobial chemicals. This whole mixture is then gradually jolted and jostled along the canal, picking up dirt and microscopic organisms along the way, steadily transported toward the outside by the movements of the jaw when we eat or talk. For humans, this movement is about as fast as our fingernails grow, but despite its snail's pace, there's very little chance that earwax will settle in and get stuck. That's because earwax, like ketchup, is what is called a non-Newtonian fluid. It's viscous until it's agitated; then it flows smoothly.

Did You Know . . . The way in which our jaw movements keep earwax moving through the ear canal is so efficient that researcher Alexis Noel at the Georgia Institute of Technology envisions it as the model for some sort of creeping filtration system to be used in robotic devices to remove dirt and debris.

And that's when I discovered that amazing seafood restaurant five years ago.

Earwax migrates the same way in sheep, rabbits, and dogs, but not, apparently, in whales. In 2007 a ship off the coast of California struck and killed a blue whale. When the body washed up onshore, local scientists went to the scene to study the remains and collect samples. One of these was a giant piece of earwax that had formed a plug in the whale's ear. This was just the beginning of one of the most offbeat yet profound scientific investigations ever.

This earplug was 25 centimeters (10 inches) long and striped, dark and light. Each stripe represented six months' time, and differences between dark and light represented times of migration and feeding. Because there were twenty-four stripes, the whale was a relatively young twelve years.

Two Baylor University researchers, Stephen Trumble and Sascha Usenko, thought that the earwax might contain a detailed record of the whale's twelve years, and analyzed it for a wide set of chemicals, including the stress hormone cortisol, testosterone, and a variety of organic pollutants.

They were right. One of the things they discovered was that the whale's cortisol levels climbed gradually throughout its life but peaked suddenly around the time it reached sexual maturity (as measured by an even bigger spike in testosterone). They suspected that entering breeding competition for the first time caused these stress hormones to rise. That explained the sudden peak, but not the steady rise through life. They wondered if the rise could have resulted from mating stress, migration, changes in food availability, or even pollutants. There was indeed evidence of persistent chemical pollutants in the earplug.

Did You Know . . . In humans there are two kinds of earwax. A single gene is actually responsible for the differences: One wax is brown, often dark brown in color, sticky, and has a definite odor. The other is almost gray, flaky, and very dry.

The genetic split is between East Asians and indigenous North Americans on the one hand, Europeans and Africans on the other. The East Asian/indigenous group has the dry, flaky version; Europeans and Africans, the sticky version. Apparently the gene mutation behind the split happened about 30,000 to 35,000 years ago—somewhere around two thousand generations back. That was about the time the Neanderthals died out. (Wonder what kind of earwax they had?)

It is slightly unfortunate (for some of us) that biochemistry dictates that those of us with dark, smelly earwax also tend to have smelly armpits. Those with flaky earwax don't.

That one animal provided a unique look into the life of a young blue whale, but there was more to come. The same team of scientists partnered with others to study not one whale earplug but twenty, including twelve from fin whales, four from humpbacks, and four from blue whales. The plugs provided a snapshot of whale life in the twentieth century, and one striking detail was how stress hormones rose and fell together with whaling.

In the first part of the twentieth century, as whaling became more widespread and efficient, the kill rates rose steadily. In the 1930s, 50,000 fin, humpbacks, and blue whales were slaughtered. The killing maxed out and so did the levels of cortisol at 50 percent above baseline levels. World War II put the brakes on whaling, and while the number of whales taken dropped, the whales' cortisol levels stayed pretty steady, even rising slightly. It's not clear why, although marine wartime activities might have had something to do with it. After the war, whaling picked up again and cortisol levels soared once more.

You might wonder if the killing of whales could really influence the levels of stress hormones in those who survived. After all, it's a big ocean. But the correlation was just remarkable: graphs matching whale deaths and cortisol march step-by-step together through the years. Even when the United States put in place a moratorium on whaling in the 1970s, whale cortisol levels stayed high, and the researchers showed that increasing sea surface temperature seems to have replaced whaling as a stressor.

Let me wax poetic, my ear.

This is not the only evidence that human activity affects whales. Even the relatively benign whale-watching ships have been shown to interfere with whales' normal feeding behavior in the tidal mouth of the St. Lawrence River, where fresh water from the river meets salt water from the Gulf of St. Lawrence.

Who would have thought earwax could be both an essential cleaning tool and a somber witness to whale life in the twentieth century?

Why do we close our eyes when we sneeze?

THIS IS A VERY GOOD QUESTION. Usually when you close your eyes without thinking it's to protect them . . . but from a sneeze? Can stuff that's sneezed out harm your eyes? Or is the pressure coming from the explosiveness of the sneeze dangerous? Let's start with the center-piece of this, the sneeze itself.

We can all agree on what sneezing feels like, but what exactly *is* it? It begins when smoke, perfume, or allergens like pollen irritate the lining of the nose. When that happens, the cells in the lining release chemicals, which in turn cause fluid to leak. All of this stimulates the hundreds of nerve endings in the vicinity, and the brain is soon notified. When the volume of incoming messages from the respiratory tract reaches a threshold, the "sneezing center" responds by switching on an automatic sequence of actions: inhaling a huge amount of air, closing the glottis (the opening to the lungs, also known as the vocal cords),

I'm snot feeling well.

then suddenly opening the glottis and expelling the air with force. The air carries away some of the fluid and the irritating substances that started the process in the first place.

But can that force be dangerous? You can read claims on the web that a sneeze propels air at speeds from 300 kilometers (186 miles) an hour to more than 1,000 kilometers (621 miles) an hour, which would be approaching the speed of sound, with the potential of creating a sonic boom. But those extreme speeds are fiction. Some scientists argue that 100 kilometers (62 miles) an hour is possible, but most recently, an international team of Canadian and Singaporean scientists determined that the maximum distance for the "plume" of droplets blasted out by a sneeze was a mere half of a meter (2 feet) and the speed was only 4.5 meters (about 15 feet) per second, or 10 miles per hour. "Leisurely" is the word for that.

Science _Fact!_ *There is a genetic variation in about a quarter of the population that triggers sneezing from exposure to the sun. It's called "photic" sneezing, or autosomal dominant compulsive helio-ophthalmic outburst (ACHOO) syndrome. Yeah, well . . . Photic sneezing can be dangerous if you're a pilot or a car driver emerging from a tunnel into the bright sunlight, but unfortunately there are no good ideas as to why this gene would even exist. Could it be that a reflex that normally contracts your pupils in the sun somehow gets its wires crossed and triggers a sneeze instead? And what's the value of a gene for this? Usually when a gene is maintained at a healthy rate in the population, it does something else that gives it value. So far, we haven't figured out the "helpful" function of the ACHOO gene.*

Despite that modest speed, people have been injured by sneezing. The most unfortunate case was a man who sneezed while he was brushing his teeth, causing a disc in his spine to bulge, which compressed the spinal cord and paralyzed him from the neck down. Less than two weeks later he died of a blood clot. Obviously this was a rare case. More common injuries caused by sneezing are strained back muscles and fainting.

And, indeed, you can damage your eyes as well. In the nineteenth century the *New York Times* reported that a woman sneezed so hard while on the streetcar that her eyeball popped out of its socket. This was an urban myth. For one thing the eyeball is held in place by six muscles, and they're much too strong to be torn by a sneeze. But a very strong sneeze *can* detach the retina in the eye. However, this injury is unlikely to be prevented by squeezing your eyes shut.

 **Did You Know . . .** Sneezing is important to African wild dogs. They are social, maintaining packs of eight to twelve dominated by a breeding pair, but there's a degree of democracy. Decisions to move on from resting involve a set of factors. Usually a single dog stands to initiate movement, and the status of that dog is important. Also, if it's the first such signal, it's less likely to be persuasive than if it's the second or third. And, finally, the decision to move is determined by a vote—and they vote by sneezing! Other species in the extended dog family make sneezing noises, but this is the only instance in the animal world where sneezing is used to communicate.

TRY THIS AT HOME! Robert Provine, of the University of Maryland, has experimented on himself to see if sneezing without closing our eyes is possible. He found he could sneeze with his eyes open; but when doing that, he noted, often the urge to sneeze also disappeared. It had the same effect as pinching the bridge of your nose when you feel a sneeze coming on, or even pushing up on the bottom of your nose. No one has a clue why any or all of these stop a sneeze in its tracks.

Provine also tried a "mouth sneeze," which requires you to pinch your nostrils when you feel a sneeze coming so that it is forced through your mouth. Provine says it feels normal, but if it is, that suggests that normal sneezes don't clear the nose as much as you'd think they should. He's also tried a "clenched-teeth" sneeze, which works, too, but apparently doesn't feel normal. One sneeze *not* to try is the nose sneeze, produced by sealing your lips. There's too much pressure behind the sneeze to force it safely through the narrow channels of your nostrils: you might damage your eardrums with the back pressure.

So what about the idea that closing your eyes prevents bacteria or viruses from flying out of your nose and mouth directly into your eyes? That seems really unlikely to me. Most of the force of a sneeze is outward and downward, not up toward the eyes. You're much more likely to get matter in your eyes from someone else's sneeze.

The truth is we just don't understand yet why we close our eyes when we sneeze.

What can we learn about human evolution from the game of darts?

THIS IS, PERHAPS, AN ODD QUESTION. But the truth is there's something special about the way we throw darts that sheds light on how our arms, wrists, and hands have evolved as we became modern humans.

It's all in the wrist.

If you have ever watched darts on television, you might be mildly surprised that anybody would watch it: hours of somewhat unathletic people performing the same monotonous tasks of stepping up to the line (technically the "oche"), staring at the target, and then, with a sudden flip of the forearm and wrist, landing the dart on the board with amazing accuracy. Then doing it again. To perform at the professional level, players have to be precise enough, at least theoretically, to release the dart within a couple of thousandths of a second of the same speed every time. Some expert throwers perfect a hand position that gives them a bigger time window, so more room for error, while others practice to make their timing more precise.

It's that flip of the wrist that is intriguing. Most dart players hold the dart beside their eye, then suddenly extend the forearm and the wrist along a straight line. The arm motion isn't particularly remarkable, but the wrist motion is. It's cocked back and then flipped forward, all in the same plane, not unlike a mechanical metronome. It's technically called radioulnar deviation.

Hand surgeons and anthropologists are so interested in this unique arm and wrist movement that it's been given its own label: dart-throwing motion, or DTM. The medical interest is prompted by the fact that knowledge of the precise nature of the wrist movements helps surgeons repair this incredibly complicated joint when it is injured. But anthropologists are curious about the DTM, too, because of what it might be telling us about our ancestors.

As our species has diverged over millions of years from the as-yet-undiscovered animal that was ancestral to both us and the chimps, major anatomical changes have occurred: our brains are bigger, we stand fully upright, and our hands have developed into diverse instruments for doing multiple tasks, from typing to holding tools to throwing. This doesn't mean that our hands have advanced and those of our relatives, the great apes, haven't—just that each species' hand has evolved to allow it to do what suits it best.

One of our distinct features is the so-called opposable thumb. Our thumbs are very different from those of chimps: if you look at the palm of your hand, you'll see that the tip of your thumb reaches as far as the first joint of your forefinger. A chimp's thumb would only reach partway up the palm, and that difference in positioning and reach gives the human hand more versatility. Our wrists are different, too.

For instance, orangutans and gibbons have subtle differences in their wrist bones that are thought to allow them to hang from a branch and rotate in a way that we can't. By contrast, gorillas and chimps are knuckle walkers, putting their weight on the last two knuckles of their fingers. Their wrist bones are reinforced to be able to bear that weight. As soon as we became bipedal, we didn't need that kind of support.

It's called DTM.

Anthropologists have speculated for decades that throwing stones at small prey might have been a crucial step in our evolution, not just for providing food but for developing the brain. We are predominantly a right-handed species, and the left hemisphere of the brain is responsible for the precise control of the right hand when it's throwing. But the left hemisphere is also (usually) responsible for language. Both language (whether speaking or signing) and throwing require precise sequences of muscle movements. If throwing came first, it could have equipped the brain to adapt its skill at movement sequences for language. The DTM, the dart-throwing motion, is the modern version of ancient stone throwing.

While the role of throwing in our evolution is somewhat speculative, there's no doubt that the invention of stone tools was a major landmark. This required leaps of imagination in the first place, hard as it is to imagine how striking one stone with another would immediately look like a good thing to do. But that was just the first leap: first the sharp edges of the crude flakes were used as cutting devices; then the mother stone from which they were struck became the object of attention, allowing the manufacture of bigger, more lethal tools, like spearheads. All of the maneuvers to create such pieces of stone craft depended on the agile hands and flexible wrists of the early humans, especially those wrist movements. And throwing spears demanded the same attention to using the arm and wrist together to accelerate the spear, releasing it at the appropriate time to ensure the correct trajectory, and doing so in the same repeatable and reliable fashion as needed to hit the triple twenty on the dartboard. To that end the two wrist bones closest to the forearm, the scaphoid and lunate, move very little during the dart thrower's motion, lending stability and consistency to that movement.

Dart throwing isn't the only descendant of some of the early uniquely human activities that our ancestors invented. Using a hammer, fly casting, and conducting an orchestra use the same motion. But the game of darts—throwing a sharply pointed object at a target—despite the pub location and its often paunchy participants, might just be the closest echo of our ancestral past.

If we have kneecaps, why don't we have elbow caps?

THE ANSWER IS SIMPLY: BECAUSE WE DON'T NEED THEM. But the backstory is much more entertaining than that.

If we did have elbow caps, they would be two more "sesamoid" bones to add to the collection in our bodies. Most sesamoids are about the size of a sesame seed—hence the name. The majority are attached to or even encased in a tendon and are thought to reduce the stress on the tendon as it slides over bones, as tendons must when the joint is moving.

We have several sesamoid bones in our hands. If you hold your right hand out, palm down, there are two tiny sesamoids right where your thumb joins the hand. There are often two more, one on the inside of your right

Sesamoids are the bee's knees.

forefinger, the other on the outside of the baby finger. One study suggested that if you go around your hand, the thumb always has sesamoids, the forefinger 50 percent of the time, middle finger 3 percent, ring finger 1 percent, and baby finger 50 percent. There are similar sesamoid bones in the feet, too, and some people have extras in both hands and feet.

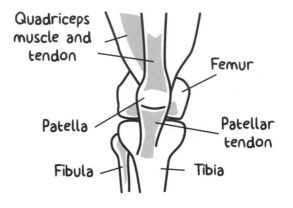

The Kick Machine

The biggest sesamoid bone in the body is the kneecap, the patella. It's a huge sesamoid, but it plays a crucial role in the movements of the knee joint. The kneecap is suspended in front of the joint, embedded in the quadriceps tendon from above and the patellar tendon below. The two tendons together help the thigh muscle, the quadriceps, to lift the knee. As the quadriceps tightens and shortens, the tendons pull and the lower leg lifts.

The patella helps this motion by increasing the angle of the patellar tendon as it connects to the lower leg bone. By raising the tendon slightly, it makes the force created by the contracting quadriceps more effective. It's been likened to the bridge on a violin, which elevates the strings and increases their tension. Also, although it's not exactly the same, it's like pushing a door open: it's much easier to do that with your hand near the knob than near the hinges. The distance away from the hinges increases the effective force. That greater effectiveness translates into better speed. In the case of the leg, a faster kick. Soccer players Megan Rapinoe and Cristiano Ronaldo have their patellas to thank.

Maximizing the force provided by the quadriceps muscle is crucial in this case because this is a joint that puts big demands on its muscles and tendons. So why don't we have an elbow cap? It is the same sort of joint, isn't it?

In some ways, yes, but not in all. Like the leg, there is a single bone in the upper arm and two in the lower, and the muscles of the upper arm, the biceps and triceps, alternately flex to bend and straighten the arm. But the triceps simply attaches straight to the prominent bony part of your elbow, with no sesamoid to facilitate things.

Part of the reason we don't need an elbow cap is that the forces involved in extending and flexing your elbow aren't nearly as great as those involved in lifting your legs to walk and run. But also, the surfaces where the bones meet are different: in the leg, the ends of the upper and lower leg bones are flat, whereas in the arm the forearm bone cups the upper arm bone, making it much more stable. If that stability contributes to more effective deployment of forces, then an elbow cap isn't really necessary. It's also true that the kneecap, by being mobile while the leg is moving, is better at enhancing the muscular force of the quadriceps than an immobile piece of cup-shaped knee bone would be.

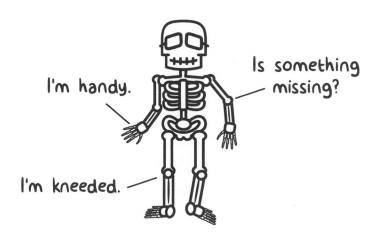

I'm handy.

Is something missing?

I'm kneeded.

Given that we are two-legged animals, it's a little surprising that quadrupeds, including all mammals, also have forelimbs and hindlimbs like ours with, for the most part, patellas on the hind legs and none on the front. But there are other curious inconsistencies: dinosaurs didn't have patellas, but very early birds living at roughly the same time did. Among modern birds, ostriches have double patellas, but other large flightless birds, like emus and cassowaries, have no patellas at all.

This puzzles me. You'd think that the large four-legged animals (bears, elk) and those big flightless birds (emus) would exert huge forces on their front legs as they ran. But no kneecap in sight. No deer, moose, bears, or horses have kneecaps in their front limbs. Do they somehow reduce the force on those legs when they run? If so, how? Or is the patella perhaps not quite as crucial as we might think?

Reptiles lack patellas, and until very recently it was thought that frogs didn't have them, either, but we now know they do—they just went unnoticed. That frogs have patellas suggest that these bones are very ancient, perhaps 400 million years old—much older than the birds living at the time of the dinosaurs—but it appears that they have appeared, disappeared, then reappeared many times in the fossil record.

As I say, the short answer is: We don't need elbow caps.

I inherited these knobby knees!

Why do beans make us fart?

It won't surprise anyone to learn that beans of any kind, whether navy, pinto, black, or kidney, are pound for pound the best generators of intestinal gas (although one 1949 study found that brussels sprouts are pretty powerful as well). But the question is: Why?

The clue lies in their advertising: they're promoted for their exceptionally high fiber content. And while that's a health benefit, it can also lead to social embarrassment, because the high fiber is largely a set of indigestible sugars—indigestible for humans, yes, but bacteria can break down these bulky molecules easily. Fortunately for us, our large intestine is home to innumerable bacteria that are capable of doing this, maintaining our general intestinal health in the process. Unfortunately for us, the side effect of their work can clear a room.

It wasn't me.

The problem is that these intestinal bacteria, your "microbiome," are operating in the absence of any appreciable amount of oxygen, so they use fermentation to break down the difficult sugars. Odorous gases are the result—sometimes in large amounts.

Science _Fact!_ *A number of studies have established that most people produce between twelve and twenty-five farts a day. You might expect that those farts, being gas, would want to rise in the intestine rather than move downward. If that were true, lying down might promote farting better than standing up. But that is not how it works: an experiment where gases were introduced to the intestine, then tracked as they were expelled, showed standing up encouraged more farting. This is because it isn't left to the physics of gases: the gut's muscular contractions force the gases out.*

In a 1991 article in the journal *Gut*, scientists described an experiment involving ten participants who had added a portion of beans to their regular diet. (A control group ate only a fiber-free diet.) All were outfitted with a "flexible gas-impermeable rubber tube" positioned in the only place it could be to collect gas from their intestines. The other end of the tube was inserted into a laminated gas bag that had been proven to be leak-free thanks to the efforts of two of the volunteers who sat with the "lower parts of their bodies in warm water for an hour," during which time no "bubbling" was observed as gas began to collect in the bags.

After the subjects' daylong gas production, the hoses and bags were removed, and all that gas was measured and tested. The bean eaters produced anywhere from just under 500 milliliters (half a liter) to nearly triple that, at 1,500 milliliters (1.5 liters). For comparison, your favorite large latte might come in close to that minimal volume, while for a liter and a half, just cross-check that with some large cartons of milk. The fiber-free volunteers, on the other hand, managed a mere 200 milliliters on average.

Perhaps not surprisingly, there is a difference between men and women when it comes to farting. And it probably isn't what you think. In another study of people who supplemented their normal diet with a ration of beans, experimenters again collected the gas produced. Besides the usual carbon dioxide, hydrogen, and methane, these included small amounts of sulfur-containing gases like hydrogen sulfide, methanethiol, and dimethylsulfide. Two odor judges (!) confirmed the significance of these sulfuric vapors for the unpleasantness of the odor, using a scale from 0 (no odor) to 8 (very offensive). Women produced much more hydrogen sulfide than men, and more intense odors, but men produced larger volumes; so, in the end, the amounts of unpleasant sulfur compounds produced by both were about equal.

Avoiding gas by not eating beans would remove an important source of fiber from your diet, so if your own farts trouble you, there are some other steps you can take.

Some reports suggest that adopting a consistent diet of beans will reduce the production of gas after a couple of weeks, although that might just be a case of people not noticing they're farting rather than an actual change in the ongoing chemistry of your gut. (But if you are a little sensitive to your fart frequency, then even fooling yourself this way is better than nothing.) There is also a recipe book-load of cooking remedies, including cooking beans with dry mustard or soaking beans for

four hours, then draining and rinsing them before cooking. There are even commercial remedies, like Beano or Digesta. Beano introduces an enzyme called alpha-galactosidase that we normally lack. It works to break down the troublesome sugars while they're still in the small intestine, so that by the time they reach the large intestine they're already digested. Digesta contains alpha-galactosidase, too, but also four other enzymes.

Did You Know . . . Monsieur Joseph Pujol (1857–1945) was a fart artist, or "flatulist." He could control his farts to produce sound effects, anything from cannons to animal noises, and even music. Onstage at the Moulin Rouge in Paris, he would even play the French national anthem, "La Marseillaise," although he needed an actual wind instrument to play the notes; he just supplied the wind. But was he a flatulist really? The sounds he made sounded like farts, but they weren't produced by intestinal gas. Pujol had an amazing ability to suck air into his body, then expel it. Skilled, yes, but just not as amazing.

A less practical but more ambitious effort to prevent the unpleasantness of flatulence was tried by the team that included the odor judges. They created a cushion of activated charcoal, a potent odor captor, and installed it inside a set of "gas-tight pantaloons." When a fart-like combo of gases was introduced, the activated charcoal, positioned this way, did reduce odor dramatically, although the researchers admitted that wearing gas-tight pantaloons with an activated charcoal cushion could be "unwieldly."

The only thing that apparently hasn't been tried so far is to introduce sources of fragrant odors into the diet—but I'm sure someone is thinking of it.

Can we smell our way home?

I SUSPECT MOST PEOPLE, unless they live in a bakery, would be tempted to say "No" immediately to this question. After all, we would need to identify odors and tell which direction they were coming from, and we'd need some sort of olfactory map to compare odors to places so we would know when we're getting close to home. But there is plenty of evidence that birds manage all of this. And there is some evidence we can, too.

How good are we at discriminating among different odors? We've all heard that dogs put us to shame with their ability to smell things like drugs in suitcases or the odors of certain diseases. But a recent study showed that we can actually distinguish as many as a trillion odors. That compares to our ability to distinguish 340,000 different tones (sound) and perhaps as many as 7.5 million colors.

Telling which direction an odor is coming from, or preferably being able to follow an odor plume as we're moving, is also within our capabilities. In one study, an irregular trail of chocolate was deposited on the ground and human volunteers were asked to crawl along the 10-meter (33-foot) chocolate path with blindfolds on.

Admittedly, a chocolate trail on the ground is a much more permanent odor source than odors of chocolate wafting through the air, but at least this experiment demonstrated that humans can follow a scent trail. And the difficulties they encountered when one nostril was blocked suggest that these men and woman were using their two nostrils to do a real-time, side-by-side comparison. If the odor faded in one nostril, they could course-correct to bring it back to both and so stay on target. And records of the paths they followed showed they did just that—exactly what a host of animals and fish do.

But a chocolate trail in the grass is fairly straightforward. Finding your way home requires the ability to use localized scents to know where you are, and even this has been demonstrated, admittedly on a small scale. In this experiment volunteers were invited into a room where there were small sealed containers lined up along the walls. In preparation for the experiment, only two of the containers had a scent put in them. (Three different odors were used: anise oil, sweet birch oil, and clove oil.) The room was 8.3 meters by 6.4 meters, or about 27 feet by 21 feet, and marked off, checkerboard-style, in squares 3 feet by 3 feet, about a meter by a meter, for a total of 63 squares.

SMELLS LIKE LAUNDRY!

The experimental subjects were fitted with blindfolds, earplugs, and noise-canceling headphones, and the lids of all the containers, including the two containing the oils, were removed. The subjects were walked randomly back and forth to ensure they were disoriented. The experimenters then took them to a pre-chosen square on the checkerboard and allowed them to stand there for one minute, asking them to "smell what this location smells like." They were then escorted back to their start locations and the blindfold, earplugs, and headphones were removed. The subject was asked to search for the squares they had just been standing on. After that search, successful or not, they were returned to the start location once again, had the blindfold, earplugs, and headphones replaced—as well as nose plugs this time—and asked to find the target squares once again.

The results might seem a little underwhelming: on average, subjects, when able to smell, could get closer to the spot they had been standing on and sniffing by something like 70 centimeters, a little over two feet. That's not exactly finding a hut in the woods, but remember that this was a closed space, the results were consistent, and the fact there was any effect at all is amazing: people were able to place themselves where they had been in the room by smell!

Science _Fact!_ *We all know that dogs have more sensitive noses than we do. For one thing, they possess up to 300 million odor receptors in their noses, while we only have about 6 million. And more of their brain is devoted to analyzing those odors. But we're not that inferior; in fact, we are better than dogs at identifying some smelly chemicals. Dogs are pretty smart, but our much bigger brains allow us to analyze and process odors in ways they can't. If we trained ourselves to detect odors the way we train our canine pets to do the same, we'd find that we're pretty good. That's how perfume, wine, and food experts are able to do what they do.*

If that's all there was, then finding your way by smell would remain, until further study, a curiosity. But one of the scientists in this experiment, UC Berkeley psychologist Lucia Jacobs, has sketched out a theory that gives our sense of smell at least partial credit for the evolution of our species.

Dr. Jacobs begins by asking why, of all the primates, we have a nose that protrudes, sometimes called the "external pyramid." The development of that nose appears to have occurred long after our brains started getting bigger and we were walking upright, a little over 1.5 million years ago. For almost a century the explanation has been that thinning the nasal passages by extending them outward creates an irregular flow that allows inhaled air to be warmed and humidified before entering our lungs. Warm moist air contributes to more efficient breathing, and that would be important for our wandering ancestors. There is evidence that people who live in climates where the air is already warm and moist have shorter, flatter noses, while those in colder climes have extended, thinner noses. But it's not clear yet that warming and humidifying happens as the theory requires.

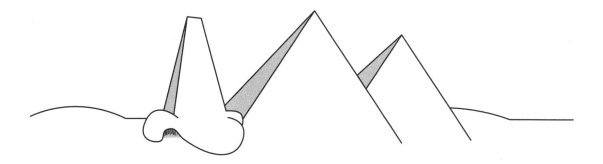

Dr. Jacobs takes the idea in a different direction. What if, she asks, the development of the modern human nose provided early humans with the ability to navigate by odor? Traces of the modern nose in fossil skulls began to appear when our ancestor *Homo erectus* started to migrate aggressively out of Africa. A wandering creature like this would need as many navigational aids as possible. It's hard to guess what the important odors might have been, but anything from fresh water to vegetation to herds of animals could have provided what Dr. Jacobs calls a "long-distance sensory highway."

The modern nose keeps the inflow of air from the two nostrils separate, possibly enhancing the ability to compare odor differences between the two, as happened in the chocolate trail experience. Another mysterious feature of the external pyramid nose is the fact that the nostrils face downward. We don't have the dog's ability to get up close and personal with smelly things on the ground (thankfully!) because we stand up to walk, so perhaps the position of our nostrils allows them to detect ground-based odors more easily.

It remains for Dr. Jacob's theory to explain why noses vary so much in shape around the world. If the nose evolved to be a navigational aid, it might be that different shapes of noses around the world are evidence that olfactory challenges changed with time and place. And it's probable that what worked best for our ancestors 1.6 million years ago was not as important half a million or a million years later, but the idea that we might have needed our noses to guide us on our first significant ancestral migration is pretty cool.

Which is more important, my big toe or my thumb?

THIS IS A FRIENDLY COMPETITION! Both are important or we wouldn't have them, and both are unique in the animal world. You might think the choice is easy, but it's important to take into account not only what each does for us today but how each enabled us to become modern humans in the first place. So what does each one do?

The big toe is the key to the way we walk. When we take a step, we literally fall forward, hit the floor with our heel, roll forward on the outside of the foot, then push off with the big toe. Estimates are that the big toe withstands 60 to 70 percent of the force of that step, which of course is why it is the biggest and strongest. Built to withstand forces, it's crucial to the fully bipedal walking that we, uniquely, possess.

I'm ready to go toe to toe.

But you can argue that the thumb is what makes us human. Period. Our thumbs are relatively long compared to the size of the hand, muscled, and articulated in a way that allows them to easily touch the tip of each of the four fingers—and exert pressure. By contrast, all the great apes have long, curved fingers, but their thumbs are much shorter than ours. Our "opposable thumb" (some apes have versions of it, too) makes possible everything from holding a hammer, a pencil, a ball, or a needle and thread to twisting the lid of a jar . . . The list is as long as the number of objects we manipulate with our hands. Some grips require power, some precision, and the thumb is crucial to all. And where we would be if we couldn't give a thumbs-up to something we like?

Science _Fact!_ *While the claim that we "evolved from the apes" is undeniably wrong, there are still some who persist in thinking that humans evolved from animals that look like modern chimps: knuckle-walking, tree-climbing, stocky versions of ourselves. Yes, we're related to chimps—they're our closest relatives among the apes—but we didn't evolve from them. We share a common ancestor, perhaps 5 million years ago, but since then we've been on different paths.*

The recent discovery of a skeleton of one of our ancient predecessors gives us some clues about the evolution of both our toes and our thumbs. *Ardipithecus ramidus*, or "Ardi," as she's called (and she was a female), lived about 4,400,000 years ago in what is now Ethiopia, making her nearly complete skeleton (an exceedingly rare find) the oldest ever discovered. From it we can tell that she was four feet tall and weighed about 120 pounds. She had a chimpanzee-sized brain and apparently walked upright—but not exactly as we do.

Her feet were adapted for both walking and climbing. Her foot was stiffened to provide the lever-like support we need to be able to stride, and her toes were capable of bending upward as they do at the end of a full step forward. But her big toe, so important to our bipedal gait, had not adapted as much to walking: it was still flexible, able to reach across to the other toes, presumably to grasp branches and move easily in the trees. It was, to some extent, an "opposable toe." So Ardi appears to have been versatile, both walking and climbing, but likely not extremely proficient at either.

Ardi's hand was not yet fully human, either, but was definitely not chimp-like: chimps' hands are reinforced and inflexible to allow knuckle walking, but Ardi was not built that way and likely didn't move around on her knuckles when on the ground. (Besides, she had feet capable of full two-legged locomotion.) Ardi's thumb wasn't quite as long as a human thumb, but it was longer than a gorilla's and even longer than a chimp's. If she's an ancestor of ours, that means that our thumbs haven't changed all that much since her time, while our big toes have done so steadily, becoming specialized for walking.

Ardi's hands and feet seem to suggest that, since our predecessors could walk upright without a fully-developed big toe, the thumb was more important in our development. Eventually, chasing game required rapid bipedal running facilitated by the big toe, but the ability to make stone tools with those versatile hands would have been even more essential to hunting. There's no evidence that Ardi made tools (that came much later) but the key question is: Did tool making accelerate the development of the human hand, or result from it? How did freeing the hands by walking on two legs change those hands?

And certainly today we see evidence that the thumb is the digit we can't do without. People who have had a big toe amputated are still capable of relatively normal walking, whereas those who

You know Ardi? We're related through marriage.

have lost a thumb are definitely disadvantaged. In fact, surgery to replace a lost thumb with a big toe is not uncommon, although highly complex. The new digit is called a "thoe."

The fact that no one would ever think to do the reverse operation pretty much seals the thumb's status as the more important body part. But I think we'd all agree that it's good to have both.

Did You Know . . . While it's true that you can lose your big toe and still walk, that's not to say you wouldn't miss it. And apparently that was true for a woman who lived about 1000 BCE in ancient Egypt. When her mummified body was unearthed during an archaeological dig, a wooden big toe was found laced to the front of her foot. It looked just like a real toe, even having a nail carved into it, and wear on the bottom showed that the woman had definitely worn it while walking.

Why do our fingertips get wrinkly in water?

WE'VE ALL EXPERIENCED THIS, whether it's too long in the bathtub or hot tub or, much less likely in these days of dishwashers, washing the dishes. The tips of fingers wrinkle significantly and only smooth out again after they've been dry for a while.

For a long time the explanation for this was uncontroversial. Given enough time, water gradually seeps into the uppermost layer of skin in your fingers but not below that, and because the top layer swells but can't separate from the layer below, it buckles and wrinkles. One researcher likened the development of wrinkles to the folding of a tent when it collapses—although the tent folds randomly. The wrinkles on your fingertips, however, line up with the ridges of your fingerprints, where the skin is anchored. The

skin collapses in between those ridges. Only when the skin dries out does it return to its normal size and fit snugly on the layer below.

It doesn't always happen to the same degree. The hotter and the more alkaline the water, the more pronounced the wrinkling, which explains why hand washing the dishes (hot, soapy water) produces this effect so well. Salty skin enhances the process, but oily skin inhibits it. We secrete oil over most of our body but not on the palms of our hands and the soles of our feet, which explains why we see the wrinkling on our fingertips and, if you think to look, your toes.

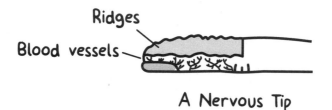

A Nervous Tip

All this seemed to be a straightforward piece of body chemistry, but it's now understood that there's also a crucial connection to our nervous system. Years ago doctors noticed that the fingertips of those who had lost sensation in their fingers through injury, like carpal tunnel syndrome, wouldn't wrinkle in hot water. That's because wrinkling occurs at the surface layers of the skin, but if the blood vessels in the layers just below are fully dilated and the pressure is high, the skin is too tense to allow wrinkling. The opening or closing of the underlying blood vessels is controlled by nearby nerves; if they are not functioning properly, the blood vessels won't constrict and wrinkling won't occur. (This reaction is consistent enough that it is now used as a simple test for nerve damage in the fingers.)

So we've known for a while how the wrinkles form, but it wasn't until 2011 that we had a theory about why fingers might have developed this way. A scientist named Mark Changizi and his colleagues argued that fingertip wrinkles allowed humans—that is, our primate ancestors— a better grip in rainy conditions, which was of huge importance for primates who spent much of their time climbing trees. Our fingertip wrinkles, Changizi points out, are actually analogous to the treads on car tires designed to give them better grip on wet roads. Where a tire comes in contact with the road, water is squeezed out from the grooves and flows away. In the same way, when a finger presses down on a wet surface, water is channeled away through the fingertip wrinkles. As the finger presses harder, more and more of the fingertip is flush with the underlying surface. The result is a firmer grip. In fact, these temporary wrinkles are better than shoes with permanent treads for secure footing on wet surfaces, because in the shoes only the raised parts of the tread, never the channels, are ever in contact with the ground.

Changizi backed up his idea by showing that the detailed patterns of wrinkles on fingertips resemble those of drainage channels found on elevated land, where surface water flows away from peaks toward valleys below. His speculative take on fingertip wrinkles didn't go unnoticed—or unchallenged. First a team from Newcastle University in the UK tested the idea by having volunteers pick up either wet or dry marbles with their right hands, pass them through a 5-centimeter (2-inch) opening to their left hands, then drop them into a box with a hole in the lid of the same width. The volunteers did this with fingers either wrinkled or smooth. When the marbles were dry, wrinkling had no effect, but when they were wet, wrinkled fingers were more adept at the task, enabling faster transfer of the marbles.

However, a second research group performed essentially the same experiment and found that wrinkled fingers were actually a disadvantage. In addition, they tested the touch sensitivity of fingertips and found no difference between wrinkled and smooth. They did admit that transferring marbles from one container to another might not be the ideal test of Changizi's theory, suggesting that measuring the force needed to pull an object out of the grip of wrinkled fingers might be more appropriate.

But Mark Changizi goes further, arguing that using marbles misses the point entirely. He'd like to see a full parkour-like test with wrinkled-fingered athletes vaulting and grabbing their way from one apparatus to another, in wet and dry conditions. Also, he argues, why not look at other species, especially those that live in wet environments, to see if they have the wrinkled-finger adaptation? Clearly, we have a lot of work yet to do!

Can we live for two hundred years?

I THINK IF MOST PEOPLE were offered a healthy, happy, and long life, they'd take it. I suspect those same people could pretty much agree on what "healthy" and "happy" would mean, but "a long life"? How long is that exactly? There are scientists and futurists today who think that the long lives that we've become used to are nothing compared to what we might be able to achieve.

Even though centuries ago few people made it to into their eighties and nineties, it wasn't because the human species was incapable of it. Infectious disease, heart disease, cancer, and stroke killed most (but not all) before they could get there. But things have changed incredibly. A hundred years ago in North America the number of years you were likely to live at birth was roughly sixty, a little more for women, a little less for men. Now it's in the low eighties, again a few years more for women. Across the developed world the rule of thumb is that for every four years that pass, we gain one year in life expectancy, and that

trend has been in place since the mid-1800s. Although that's obviously a broad average, if you break it down, it means that a baby born right now might live five minutes longer than one born twenty minutes ago. I hope they enjoy those five minutes!

Much of that improvement has come with reduced infant mortality, better treatments for infections, like antibiotics, and preventing death from heart disease. But is there an upper limit?

The ten oldest women ever were all more than 116 years of age and the oldest (although there is now some doubt about her documentation) was Jeanne Calment of France, who was over 122 when she died in 1997. There are several women living today who are 116. The ten oldest men ever didn't live nearly as long: only the oldest, Jiroemon Kimura, lived to be 116. The oldest men today are around 112.

Those are impressive numbers, but can we live longer than that? A growing number of scientists think so; some of the boldest argue we can live much, much longer. What gives them that confidence?

Well, many animals live many more years than we do: rockfish can live to be 150, bowhead whales to 200, and the amazing Greenland shark to 300. But they're not the most intriguing. There are animals living today that actually seem to be immortal or so long-lived as to be off the charts. They are physiologically very different from us, but they may hold some chemical secrets to longevity.

Science Fact! Bowhead whales live in the western Arctic and are known to live at least two hundred years. One of the ways we know this is that dead bowheads have been found with harpoons embedded in their blubber—harpoons from the late 1800s!

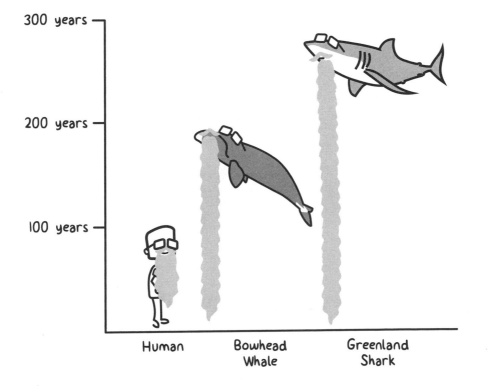

One is the hydra, a one-centimeter ambush predator that lives in ponds. It has a tube-shaped body with tentacles at one end; the other end anchors to a plant and waits for prey to swim by. They're amazing animals: they can be cut in half, cut into quarters, sliced and diced this way and that, and each piece will grow a new hydra. The reason they never seem to die is that they have an inexhaustible supply of stem cells.

Stem cells are cells that can transform into specialized cells and help repair adult tissue. Unfortunately, while we're born with ample supplies of stem cells, most of them are recruited to become other types of cells as we mature and—unlike the hydra's—relatively few are the sort that help with regeneration. And, sadly, we mostly don't replace them.

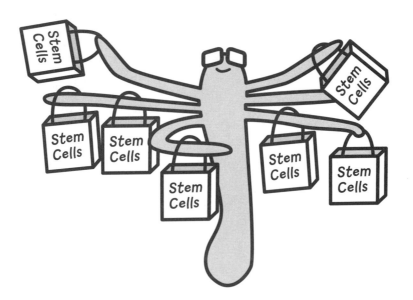

Naked mole rats are another bizarre example. They are small mammals that live in underground colonies in eastern Africa, and while they're slightly bigger than mice, they live not just about six years, as you'd expect, but more than thirty. Because they haven't been studied that long, no one knows exactly how long they can live.

Also puzzling is that they have very high levels of free radicals, chemicals known to cause tissue damage and are assumed to play a role in aging. But they don't seem to matter to naked mole rats.

If we could find a way to continue to increase our stem cell count, like the hydra, or resist free radicals, like naked mole rats, we might be able to get past 115 years, which may be a good rough estimate of our maximum age. Biomedical gerontologist Aubrey de Grey has identified what he thinks are the seven crucial issues we need to address if we are going to stop aging. He argues that if we deal with each one until they're all solved, the sky is the limit, and that stem cells, like those keeping the hydra alive, will play a crucial role.

Another optimist, futurist and inventor Ray Kurzweil, believes that within a couple of decades we will have invented "nanobots," microscopic robots that will patrol our bodies and repair the damage of aging as it happens (see page 165). Once we have those, he says, there's nothing to stop us living well past two hundred.

Did You Know . . . In case you think that Aubrey de Grey and Ray Kurzweil are just dreaming, it's worthwhile knowing that Google has started a company called Calico (California Life Company) that is dedicated to the business of life extension. It has recruited some of the very best experts in aging to their roster of scientists.

Of course there's one unanswered and very important question: if you were to live to two hundred, what sort of shape would you be in? If you spent the last hundred years of your life as a very aged *but still living* person, that might not be what you were hoping for. On the other hand, if all life events were just stretched out, so you'd be a teenager for fourteen years instead of seven, and middle age would happen around age one hundred, that might not be so bad!

History Mystery

What really caused the mass hysteria during the Salem witch trials?

THE STORY OF THE TRIAL and execution of the Salem witches has been told and retold. From the fall of 1692 to early winter 1693, nineteen so-called witches were hung, and one was crushed to death under a pile of stones. Today the "evidence" against these men and women would simply be laughed at.

But one of the lingering puzzles is the behavior of the (mostly) young women and girls who claimed to have been afflicted by the "witches." Most explanations for their actions come down to some form of mass hysteria, but scientists have attempted to find evidence that disease or poisoning might have caused them to do the things they did. Each theory has intriguing possibilities but also flaws.

There's no doubt that in the early 1690s, in Salem, there was an epidemic of weird, even inhuman conduct. Contemporary accounts describe how both boys and girls complained of seeing horrible visions and feeling as if they were being pricked with pins. Children were temporarily blinded, their eyes rolled in their heads, and they barked like dogs. They also screamed in pain and inexplicably ran and jumped, then lay completely still with their limbs at odd angles. It's not surprising that folk remedies like cutting locks of hair from a suffering child, frying it in a pan, then throwing it on the floor didn't help. In this very religious society, doctors, amazed and completely befuddled by fits and spasms the likes of which they had never seen before, soon concluded that Satan and his witches were responsible.

Today's sociocultural explanations for the accusations and subsequent trials tend to focus on the accused. They were adults, often outcasts in Salem society even before the allegations of witchcraft, or were in conflict with people

in the community, suggesting that the charges were used to settle personal scores. These explanations also focus on seventeenth-century Christianity's belief that women were more susceptible to Satan than men. The scientific explanations focus instead on the symptoms of the "victims."

Science _Fiction_! _While most of the accused were women, men were also targeted. Of the twenty executed for witchcraft, fifteen were women and five were men. One of those men was a forty-two-year-old minister who was suspected of being the mastermind behind the spells. Satan's power seemed to have no boundaries: two dogs were executed for being witches, too._

In 1986, Linnda Caporael at Rensselaer Polytechnic Institute argued that the children displaying the terrifying effects of "witchcraft" were actually poisoned by ergot, a fungus that infects rye.

If the weather is damp, the fungus can infest the crop with its spores, and any product made from that rye, like bread, would be contaminated. Ergot spores contain a variety of toxic drugs, including a relative of LSD called lysergic acid amide, giving it 10 percent of the hallucinatory potential of LSD. It can produce several symptoms reminiscent of those of the accused in Salem: unusual spasms and maniacal excitement, as well as hallucinations and delusions.

Caporael went on to argue there was good reason to believe that ergot could have contaminated the rye crop that

Wanna take a trip?

year in Salem, because the fungus had been noticed in and around Salem before then, and that spring had had the warm, wet weather necessary for ergot to spread. A man named Thomas Putnam had the kind of wet lowland field where ergot could flourish; three of the original accusers were members of his family. He also likely gave some of his rye stores to the local minister, Samuel Parris, who was often partially paid with provisions like grain. Parris's two daughters also succumbed to the mystery condition.

Caporael's claims were almost immediately disputed by scientists Nicholas P. Spanos and Jack Gottlieb. They offered numerous rebuttals, the most damning of which might have been that close scrutiny of accounts of the "bewitched" girls' behavior suggested that they were reacting to cues. If an accused witch glanced at them, they fell down; if she bit her lips, they shouted that they were being bitten. Most of their actions and complaints fit with what was expected of bewitchment at the time and so suggest they were consciously acting out a role, not behaving as if drugged.

A third author, Mary Matossian, a historian at the University of Maryland, weighed in, arguing that many domestic animals were behaving strangely at the time, and they could hardly be reacting to human cues. Also, of course, if the rye were contaminated, they would have been eating it, too.

Today the discussion of ergot poisoning has ground to a halt, but theories of what happened at Salem have not. An independent scholar, Laurie Winn Carlson, published a book called *A Fever in Salem: A New Interpretation of the New England Witch Trials*, in which she claimed the cause to be something called encephalitis lethargica, an epidemic of which struck Europe at the end of World War I, between 1918 and 1920. Half a million people were killed by the disease, but what's the evidence it affected the Salem residents more than two hundred years earlier? Many of the symptoms seen in the "bewitched" in Salem were echoed in the 1918 version: pain in the limbs, partial paralysis of the eye muscles, panic, hallucinations, tongue biting, and weird body movements. Even the cued responses seemed to happen: patients in 1918 would become agitated if their name was mentioned, and young people, more of them girls than boys, were affected—again, as in Salem. However, there were many deaths in the 1918 epidemic, and the only deaths in Salem were at the gallows.

Identifying the cause of encephalitis is tricky. Most kinds of encephalitis are produced by viruses, but there is only slight evidence that the 1918 version was caused by such a virus. But even if the Salem outbreak had the same root cause as the 1918 epidemic, why did it strike Salem and not, say, Boston?

Even an encephalitis first discovered in 2007 has been suggested as a suspect in Salem. It involves an autoimmune attack (the immune system mistakenly

targeting the body's own tissues) on neurons in the brain. The symptoms are amazingly similar to those in Salem, but as usual there are pieces that don't fit, like the fact that this encephalitis is associated with tumors 50 percent of the time. No evidence of that in Salem.

So that's where we stand today. There have been several pretty convincing theories for the behavior of the children in Salem more than three hundred years ago, but each faces contradictory evidence, seems insufficient, or depends on unlikely circumstances. This "history mystery" remains unsolved.

Part 2
Mystifying Animals

How do homing pigeons find their way home?

THESE DAYS WE CAN GET IN THE CAR and a voice will guide us accurately anywhere we want to go. Pretty amazing.

Well, not really.

Amazing is the fact that homing pigeons, a variety of the plain old city pigeon, whose brains are about the size of the tip of your index finger, can find their way back to their home roosts from hundreds of kilometers away. A remarkable feat, especially since, once they're released, after a quick circle or two, they fly more or less in a straight line back home.

Pigeons, despite those tiny brains, use an array of signals to guide them home. One is the sun, but it's complicated. As it tracks across the sky from dawn to dusk, its position changes according to the time of day: east in the morning, south at noon, and west at sunset. So, to use the sun to direct you, you have to know what time it is.

Can pigeons tell time? Yes, it turns out, they can! Researchers confirmed this by keeping pigeons in rooms with lights simulating day and night—but offset from real time by several hours. Those pigeons, when they were released outdoors, tried to fly home but headed in the wrong direction—the direction you'd expect if their clocks were off by the same amount that the lab had shifted their "daytime." So it seems they do have internal clocks that allow them to use the sun as one of their homing tools.

Another tool is likely the Earth's magnetic field. When we use a compass to navigate, we are using the Earth's magnetic field to guide us. The compass needle points to magnetic north, where the Earth's magnetic lines of force converge.

Scientists began to think that pigeons might have their own internal compasses when they noticed that some of them released near a magnetic anomaly, a place where the Earth's magnetic lines are deflected by the metallic or chemical constituents of the local rocks, flew off in the wrong direction. Fitting tiny magnets on pigeons' heads caused the same misdirection, and recent experiments have shown that the birds are sensitive not only to the direction of the magnetic lines of force but also to their angle. (Magnetic lines are horizontal at the equator but angle downward toward the poles.)

But how do the pigeons perceive that magnetism? They do have tiny bits of magnetic material in their beaks, but it is doubtful that's what they're using. Some experiments have suggested that they can "see" the Earth's magnetic lines by virtue of complex reactions happening in the retinas of their eyes. But however they do it, magnetism is part of their navigational repertoire.

You never ask for directions!

Darling. I know the magnetic lines like the back of my wing.

Science Fact! *Magnetic north is not the same as the North Pole. These days they're about 500 kilometers (311 miles) apart, and the magnetic north pole is continuing to move farther north at more than 50 kilometers (31 miles) a year. At that rate, it will one day leave Canada and end up in Siberia.*

And so are odors! One of the coolest experiments with homing pigeons was conducted in Germany and Italy. Two sets of pigeons were transported across the country, in opposite directions, in vans equipped with activated charcoal filters, so the birds couldn't smell the air during their travels. The birds were then allowed three hours of breathing fresh local air at their different destinations. Then one group of pigeons were again sealed off from the outside air and driven across country, about 50 kilometers (31 miles), to join the other group. So there were two groups of pigeons in the same place, but only one had been exposed to the local air. When released, the two groups headed off in different directions, probably because the much-traveled group assumed they were still at the place where they had been able to smell the local air.

This incredible ability probably isn't dependent on a single scent but likely a combination of two or three, each varying in different directions. If there's a little more of this odor, a little less of that, then the bird compares that mix to what they smell at their loft and follows the familiar combination home.

Did You Know . . . It's not just homing pigeons that use odors to find their way around. Some seabirds, like albatrosses, forage for food over thousands of kilometers of open ocean, all of which looks pretty much the same. Yet they're able to locate schools of fish by using their very prominent olfactory bulb, that part of the brain devoted to identifying odors, to smell their prey from 20 kilometers (about 12.5 miles) away. They find half the food they eat this way!

If this all weren't enough, homing pigeons apparently also use infrasound to navigate.

In the 1960s pigeon researchers at Cornell University were puzzled. Their homing pigeons had no problem finding their way home except when they were released from three particular places. From Weedsport and Castor Hill, New York, they always took a wrong turn as soon as they were aloft, and from Jersey Hill they flew in random directions, except on one day, August 13, 1969, when the Jersey Hill birds flew directly home. For decades no one had a clue what was going on.

Then a geologist named Jon Hagstrum wondered if infrasound might be involved. Infrasound is an extremely long-wave, low-frequency sound, far below our ability to hear. Many animals, like whales, elephants, and even tigers, generate infrasound and communicate with it. Pigeons don't make infrasound, but Hagstrum wondered if they might use the infrasound produced by ocean waves pounding on the shore, since it can travel thousands of kilometers in the earth and through the atmosphere.

Home is that way.

Don't be feather-brained! It's this way.

Weedsport's Infrasound Crisis

He was able to gather precise weather data for August 13, 1969, and discovered to his amazement that on that day the wind direction and speed were perfectly aligned to allow an infrasound signal, usually channeled high up into the atmosphere, to be guided directly to Jersey Hill. For this one day the pigeons could find their way home. At Weedsport and Castor Hill, infrasound is almost always diverted both by the local landscape and prevailing winds so that it arrives from the "wrong" direction. Apparently, the pigeons are misguided as a result.

And finally, pigeons also likely use sight to get home. Once their various tools have brought them into the vicinity of their loft, pigeons use visual cues to get them all the way home.

Why do moths fly to lights?

"LIKE A MOTH TO A FLAME" describes how we all can be attracted to something that isn't good for us. A version of this saying has been around at least since Shakespeare's time, and it is based on a natural phenomenon.

Nineteenth-century entomologist Jean-Henri Fabre described this vividly. One morning Fabre put a huge female great peacock moth in a cage and that night, to his amazement, discovered

You light me up.   That burns.

that at least forty male moths had flown in the open window and gathered around the cage. Today we understand that the female was emitting pheromones, airborne chemical attractants, and the males had detected them and rushed to her side.

But part of this event still defies comprehension. When Fabre brought a candle into the room to better see what was going on, the male moths flew at the flame, sometimes "putting it out with a stroke of their wings." It made sense that male moths were attracted to the female; it made no sense that they would be attracted to a flame, especially since the female was right there.

The modern version is, of course, moths flying to the porch light or a streetlamp on a summer night. Many will fly continually around the light, but if there's a nearby surface to land on, many settle. But why do they do this? The vast majority of moths are nocturnal (likely to avoid avian predators), so, if anything, they should avoid lights. And in fact their light attraction can be fatal, as bats quickly figure out that outdoor lights are a good place to find insect prey.

One popular theory is that moths mistake these lights for the moon, as on a moonless night streetlights are the next best thing. The idea is that moths that navigate great distances need some sort of beacon to keep them on track. The moon would be perfect, at least on clear nights, although a moth would require some sort of internal clock as well. In the Northern Hemisphere, as the night wears on, the moon crosses the sky from east to west. Keeping the moon to its left would mean the insect would first fly south, then west, and finally north. A clock would help it compensate for that movement and fly in a straight line.

This theory has been extended to explain the details of a moth's behavior around a light. The claim is that as the moth gets close, it has to fly in tighter and tighter circles (to maintain its position vis-à-vis the light) until it spirals right into it. But a few minutes' observation at an outdoor light in the summer shows it's not that simple. Some moths do indeed hit the light, but others just circle it at a distance, and yet others land nearby and remain motionless.

I wish I could use my flashlight.

So what are moths actually doing? The problem is it's hard to follow a flying moth in the dark for more than a few meters. Of course, if you use a flashlight, you'd likely attract it. But scientists are nothing if not resourceful.

Dr. Robin Baker is a good example. In the 1970s, Dr. Baker, then at the University of Manchester, glued cotton threads to the backs of moths, then tethered the threads to a jointed mobile arm attached to a recording apparatus, so that no matter which way the flapping insect turned, a record of its movements was preserved. (It would be like running on a treadmill mounted on a turntable: you could turn in any direction you like.)

In this experiment the moths Baker were testing did seem to use the moon as a guide, but, strangely, they did not adjust to its movement across the sky through the night, meaning that if they were flying free they would be changing direction continually. On the other hand, when the moon was obscured by a screen, the moths altered their flight pattern abruptly and began to fly in random directions. That scored points for the moon idea. Also, tethered moths were more attracted to artificial lights if the lights were about the height of a three-story building instead of about knee-high and were about the size the moon would look to a moth. This again suggested the moths might be mistaking those lights for the moon.

Baker did point out that his experiment would only work for moths intent on flying long distances, such as those that are migrating. Not all moths do.

If you think that tethering moths to a cotton thread and allowing them to fly is a little odd, what about gluing them to a boat? Yes, that's been done, too. Dr. H. S. Hsiao of the University of North Carolina designed this setup. His moths were glued to little Styrofoam boats in a miniature pool with a light at one end. When the light was on, the moths approached it all right, but they didn't take a spiral path. Instead they flapped their wings, dragging their little boats with them toward it. Some banged right into the light, but most aimed for a spot to one side or the other—a puzzling result. At no time did they appear to spiral into the light, and some even appeared to turn away from the light at the last moment, prompting Dr. Hsiao to suspect that, in the end, they were—oddly—trying to avoid the light.

One more complication here is that there are moths that aren't attracted to streetlights at all, which might hold significant advantages for them. Moths that are attracted to outdoor lights spend less time feeding, which affects their health, especially their ability to reproduce. As cities light up more and more around the world, their populations may decline while those of their light-avoiding cousins grow.

All the experiments taken together suggest that lighted objects, especially at about the apparent height and size of the moon, might attract moths. But remember Jean-Henri Fabre's moths attracted to his candle flame? That candle was on a table, not high in the sky, nor was it round like the moon. No, it appears that the attraction of moths to flames will remain an enduring mystery for now.

What is a living fossil?

A "LIVING FOSSIL" SOUNDS LIKE AN OXYMORON, doesn't it? Fossils, after all, are the remains of flora and fauna captured in river sediments or tar pits, usually millions of years ago. Flora and fauna that are now extinct, right? Well, not so fast.

Of course, many extinct species can be linked to living ones. The modern horse, for example, was preceded by at least a dozen related species. But those earlier species appeared and then

I'm 400 million years old.

You're a fossil.

disappeared over the course of 50 million years. Nonetheless, there are some plants and animals that appear not to have undergone the same kind of evolutionary journey. In fact, Charles Darwin came up with the idea of "living fossils" to describe animals or plants that seem to have changed little, if at all, in millions of years.

Take the coelacanth.

There are plenty of fossils of this huge fish-that-looks-like-it-wants-to-be-a-four-legged-animal; with fins like paddles, it can be as long as 2 meters (6.5 feet) and weigh up to 90 kilos (nearly 200 pounds). The oldest coelacanth fossil is about 400 million years old, evidence that this fish lived alongside those animals that actually *did* become the first to venture onto land.

The speculation was that coelacanths died out around the time of the extinction of the dinosaurs, 66 million years ago. That is, until a live coelacanth, looking exactly like its fossilized ancestors, was discovered in a fisherman's catch in South Africa in 1938.

We now know that coelacanths still live along the coasts in the Indian Ocean and Indonesia, but at pretty good depth: 700 meters (nearly 2,300 feet). There might only be several hundred alive today, but that's several hundred more than we thought there were, earning them the title "living fossils."

But how does an animal like the coelacanth manage to remain unchanged when most living creatures have had to evolve to survive? Some argue that its habitat—deep waters near coastlines—is relatively safe from predation. Also, they smell terrible and taste awful, perhaps making them an undesirable supper and allowing them to endure.

Something smells fishy.

But many evolutionary scientists argue that this whole idea of a "living fossil" makes no sense. They claim that coelacanths *are* changing and *have always been* changing; it's just that the changes are slow and subtle. Actually, there have been various shapes and sizes of coelacanths over the millennia. What's more, modern science has been able to identify changes in the coelacanth genome, most of which are not apparent when you look at the fish itself. So while the living fossil idea comes from fish or animals or plants simply *looking like* their fossil relatives, it may be a bit of a misnomer.

Science Fiction! *The term "missing link" was invented to describe an extinct animal that would fit perfectly into a chain leading from an ancient species to a modern one. It was often applied to human evolution. A new fossil of an early animal would be discovered with some apelike qualities (like the ability to climb) and some human similarities (a skull more human-shaped than ape) and would immediately be labeled the missing link. But it's now clear that human evolution doesn't work that way: instead there are all kinds of species somehow involved in our past. For instance, we now know that just about everyone alive today contains some genes from the Neanderthals, which until recently were thought to be a completely distinct species that had nothing to do with us.*

The ginkgo tree is another example of a living fossil that isn't quite that. It has an ancient lineage, going back more than 200 million years, to a time when cockroaches were 90 percent of the world's insects, giant dragonflies patrolled the air, and reptiles (not yet the dinosaurs) ruled the land. While there is only one species of ginkgo alive today, there were many more in the past. However, despite those extinctions, it's pretty certain that today's tree has been around for more than 100 million years, unchanged at least in appearance. But, again, there's that catch: ginkgo leaves look just like they did millions of years ago, but the ginkgo genome suggests there has been constant genetic change over that time.

It is likely that the genetic differences have helped the tree survive: many of the ginkgo's current genes serve to protect it from insects or even attract predators of those insects. The modern trees are also resistant to disease, can live more than a thousand years, and occupy a very narrow environmental niche—all characteristics that "living fossils" tend to share.

Did You Know . . . If you ever doubted the ginkgo's ability to survive when times are tough, how about this: there are six ginkgos in Hiroshima, Japan, that stand no more than 2 kilometers (just over 1 mile) from where the 1945 atomic bomb exploded. No other plants or animals around them survived. The six trees are alive and well.

There are many other things that have been labeled as living fossils, like the tuatara, a reptile native to New Zealand, and the duck-billed platypus in Australia. But while they all look much like their ancient ancestors, their genomes reveal they haven't stopped evolving. (In fact, the tuatara is changing genetically faster than any other animal that's ever been analyzed.) Yet, somehow, in each of these cases, those shifting genomes and the specific environments in which the animals live have conspired to allow these select species to maintain a consistent appearance over vast stretches of evolutionary time—making them appear to be "living fossils."

How do squirrels find the nuts they bury?

SQUIRRELS BURY A LOT OF NUTS. It's impossible to estimate exactly how many, but wildlife experts have suggested that a safe guess is hundreds or even thousands each fall—per squirrel. Some squirrels, like red squirrels, create one huge cache for all of their nuts. But gray squirrels seem to love a real challenge, because they "scatter horde," hiding a single nut at a time, each one in a different place. So how do they find them when they need them?

If you'd asked scientists this question a hundred years ago, most would likely have pointed to the squirrel's sense of smell. It's true that their olfactory ability is more acute than ours, and that they could likely detect an acorn under a few inches of snow plus a couple of centimeters (an inch or so) of earth, but that wasn't the main reason that smell was thought to be the key. Instead, scientists back then just couldn't believe that squirrels had a big enough brain to remember where all those nuts were hidden.

Let me give you the nuts and bolts.

Today the thinking has changed. Recent research has shown that squirrels can remember where they've buried most of their nuts.

One of the first experiments to prove this was done in an outdoor arena a little less than 45 square meters (just under 500 square feet). Squirrels were given hazelnuts to eat until their hunger was satisfied, then given more. The first squirrel was let loose on its own in the arena until it had buried ten of those extra nuts. After the squirrel was returned to its cage, the experimenters snuck back into the arena, removed the nuts, and filled in the holes. The next squirrel was then released into the arena with its surplus nuts, and the routine was repeated. By the time all the squirrels had had a turn, there were emptied and smoothed-over cache sites all over the arena. Then the scientists returned to the arena and put fresh nuts back in the first squirrel's hiding places—but added an additional ten nuts in other squirrel's holes. The first squirrel was allowed back into the arena to dig up ten nuts.

If memory was the most important cue, a squirrel would tend to dig up nuts from the holes it had dug itself; if smell were more important, it would dig up whatever it smelled first, which wouldn't necessarily be from its own hiding spots. Memory turned out to be the key: each squirrel dug up significantly more nuts from the holes it had created. And these tests were given two, four, and even twelve days after the nuts were buried, showing that the memories were definitely not fleeting.

Of course, this was an experiment in a defined space over a relatively short period of time with tame squirrels, so it isn't a complete picture of what happens in the wild. *That* process requires multiple decisions, which, over the course of a hard winter, might be the difference between life and death.

A squirrel picks up a nut. Before it even makes a move to bury it, it has to decide whether the nut is worth burying. It therefore makes a series of judgments using two simple movements: handling the nut in its paws or flicking its head with the nut in its mouth. By using a combination of these, it can determine the weight and whether the shell is intact and free of insect infestation. (Squirrels tend to eat acorns with insects right away instead of caching them.) All these are predictors of how long the nut will last underground.

Did You Know . . . In eastern North America, squirrels eat acorns from both white oaks and red oaks, and they know the difference based on shape and chemistry. Red oak acorns are dormant over the winter, but white oak acorns have already germinated by the time the squirrel harvests them and will likely rot underground. Sure enough, squirrels tend to eat the white oak versions right away but cache the long-lasting red oak acorns.

Once the squirrel is satisfied that the acorn is good to go, it will head out to bury it. But first the squirrel has one more thing to consider. Burying the nut nearby raises the risk that another squirrel might happen upon it by chance or by design.

There are always squirrels with overlapping territories close to a grove of nut-producing trees. Given that they are always watching, there's a good chance that a squirrel will be observed when it caches a nut. That's the advantage of traveling farther out of the high-density area, especially when the nut is high-quality. But that carries its own risks. Scampering long distances across the white snow in the winter to recover a nut would be an advertisement to predators.

That's why some squirrels become masters of deception. They have been seen turning their backs on other squirrels when they bury a nut and also making a show of digging decoy caches and covering up the empty holes.

And finally, there are all those different kinds of nuts! They are a challenge to squirrel memory, but the rodents seem to have a technique that will help. It's called chunking—that is, grouping

similar things together. An experiment with fox squirrels on the campus at the University of California, Berkeley, showed that, given a supply of almonds, hazelnuts, walnuts, and pecans, the squirrels definitely preferred to cache nuts of the same kind together. This technique likely allows a squirrel to remember, for instance, that all the almonds are over by the rocky outcrop, but the walnuts are near the fence. Why bother? Nuts differ in their nutritional value, so in a hard winter it might be important to know where the super-nutritious hazelnuts are.

The question was "How do squirrels find the nuts they buried?" The answer is: "It's complicated."

The Nutcracker

What is kopi luwak?

It's BOTH AN AMAZING CUP OF COFFEE and a depressing but classic story of human behavior.

The animal in the crosshairs of this story is the palm civet, sometimes called a polecat, sometimes a civet cat. Found in parts of Asia, the palm civet is not a cat; it looks a little like a mink but is gray and black, at most 5 kilos (11 pounds), with a masked face reminiscent of a racoon.

These particular civets are mostly herbivorous, although they are part of the order Carnivora. They are nocturnal and arboreal. And one of their favorite foods is the coffee bean—actually the coffee cherry, the fleshy fruit around the bean. It's claimed that they seek only the ripest cherries on the tree, swallow them whole, and eventually poop the beans out.

I've bean through a lot.

That's where humans come into the story. A commonly told tale is that when the Dutch began to establish coffee plantations in

Indonesia in the eighteenth and nineteenth centuries, they wouldn't allow the local population to have coffee. People started collecting civet dung, washing the coffee beans, fermenting and sun-drying them, then roasting to make coffee for their own use. Another account suggests they sold the end product to the Dutch.

If the quality of the coffee had been more or less the same as coffee gathered before going through a palm civet, it would have simply been a story of some people who made the best of a bad situation, and ended there.

But the coffee wasn't just unbelievably good; it was *great*. With its jungle and chocolate undertones, it might even have been the best coffee ever made! That's what people said, and it became the most sought-after coffee in the world. It's hard to identify a consistent cost, but prices have ranged anywhere from US$80 a cup to US$3,000 a kilo. Today prices seem to have settled around US$500 to $600 a kilo.

We have to look to science to understand why, according to thousands who love their coffee, kopi luwak is so different and so good. What exactly happens to the coffee beans while they're inside the civet? The best guess is that the civet's digestive enzymes go to work on the proteins on the surface of the bean, catalyzing their breakdown. It's presumed that this reduces the bitterness. Dr. Massimo Marcone at the University of Guelph, in Ontario, reported seeing extensive pitting on the surfaces of the civet beans using scanning electron microscopy. He also reported the beans to be harder and more brittle than those straight from the tree—and lighter in color.

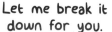

So a few dung hunters wander the forest and make pretty good money for gathering up civet poop to make a terrific cup of joe. But, sadly, the kopi luwak story doesn't end there. "Pretty good money"? It was *huge* money, and soon entrepreneurs started to farm palm civets—battery-cage them, giving them nothing to climb on, force-feeding them tons of coffee cherries, and cutting back on care and cleaning. These extremely lucrative "commercial" versions of kopi luwak coffee hit the market, and the controversy over their practices wasn't far behind.

In 2013, the BBC sent undercover buyers to Sumatra, who saw conditions that disgusted them. They made these discoveries public. That exposé was followed by a report in the journal *Animal Welfare*, which, if anything, painted a bleaker picture of the civets' confinement. At about the same time, responding to evidence that the lucrative market was encouraging fake or diluted kopi luwak, a scientific team from Japan and Indonesia published results of tests that demonstrated that authentic kopi luwak coffee could be distinguished from fake kopi luwak, a 50 percent kopi luwak blend, and other coffees. The kopi luwak bubble was bursting.

Did You Know . . . It apparently wasn't enough that we have coffee beans rescued from feces of the civet. Now there's elephant poop coffee. It's even more expensive and definitely harder to recover the beans: for every pound of so-called black ivory coffee, someone has to sort through 15 kilos (33 pounds) of elephant dung.

So today the question is: If money is absolutely no object, but conservation is, how can you satisfy yourself that the kopi luwak coffee you want to buy is produced from civets in the wild?

Some coffee companies advertise that their product is gathered the traditional way, claiming it comes from a small organic farmers' collective, or that some of the proceeds go toward education and vocational training in remote parts of Sumatra. Yet some major international coffee growers' organizations refuse to certify any kopi luwak because it's so difficult to prove where the beans have come from once they're in the package.

The team that showed they could identify 50 percent kopi luwak blends or even fake kopi luwak have yet to be able to tell natural from commercial kopi luwak; a technique like that would go a long way toward easing consumers' consciences. But in the absence of a definitive test like that, this is one cup of coffee that you should probably take a pass on.

Which animal has the most powerful bite?

THIS QUESTION IS A BIT LIKE ASKING: Who is the best athlete? Can a hockey player be compared to a golfer? What about: Pound for pound, can a featherweight boxer be compared to a heavyweight? And what about all those athletes who are no longer alive and who played at a time when the rules and equipment were much different from today? Do they count?

It might surprise you, but when comparing the biting ability of animals, there is a measure that can be used to compare them all, whether they're birds, mammals, reptiles, or fish. It's called "bite force": the amount of force an animal's teeth exert when it bites down with maximum effort. Sometimes that force can be measured directly in the living animal, sometimes it has to be estimated from fossil jaws and teeth, and sometimes it can be scaled to the animal's body size.

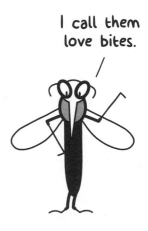

I call them love bites.

Force is measured in "newtons." Human front teeth can exert a force of about 500 newtons, but the molars at the back of the mouth can bite about twice hard. Body size is important to bite force, so a gorilla might have a bite of 1,700 newtons, and a chimp, which by body weight isn't any bigger than a human but is much stronger, will top out at about 1,500. But these numbers are nothing compared to the champions of bite.

 TRY THIS AT HOME! Want to understand what the force of one newton feels like? Rest an average-size apple on your flattened palm. The pressure you feel is approximately one newton! Not "Newton's apple" but the apple's newton.

A Single Newton

For instance, alligator and crocodile species across the globe are all-stars. The American alligator can bite up to 6,000 newtons, and one saltwater crocodile set a record for a living animal at 16,414 newtons! (The big reptiles beat out the former record holder, the hyena, whose much lesser 4,500 newtons can still crush bone.) As a testament to the importance of body size, this frightening croc weighed 3,689 pounds (almost 1,700 kilos). But an additional ten crocodilian species topped the gorilla. The researchers who gathered this data (at least partly by strapping crocs on a board and inserting a pressure gauge into their mouths) were surprised to find that it's not just that body weight is relevant; it's really the only significant variable. The shape of the snout and the diet, both of which vary significantly for these animals, just don't make much difference. It's all in the poundage.

All the crocs in this study were alive, but—given that body size is the most important determinant of bite force and that much bigger saltwater crocs have been documented—these bite forces aren't the limit. For instance, one croc measured years ago at 6.7 meters long could likely have bitten down with a force of 27,000 newtons, and who knows about even larger, extinct versions.

But for now, let's stick with 27,000 newtons. That's probably the best you can find on Earth today. But the beauty of measuring bite force is that you don't have to have the actual living and breathing animal with you there in the lab. Fossils can be computer-modeled, and the bite forces associated with them can be estimated. And that opens the door to the dinosaurs and, of course, *Tyrannosaurus rex*.

Now, this is where body size really comes into its own. The *T. rex* named Sue in the Field Museum in Chicago is estimated to have weighed 8,000 kilos. Even though estimates of her bite force vary according to the methods used to reconstruct them, one of the latest estimates has Sue's bite ranging upward to 34,500 newtons. Gregory Erickson of Florida State University, an expert on bite force, likened *T. rex*'s bite to having three small cars piled on top of its jaws.

Did You Know . . . Many attempts have been made to measure the bite forces of man's best friend, but as one group of researchers admitted, their measurements really depended on how enthusiastic the dog was! But, from a mix of actual bite measurements, estimates based on musculature and skulls, and computer modeling, the average numbers are somewhere between 1,000 and 2,000 newtons in dogs—and a mere 100 or so in cats. In the cats' defense, there were no recorded measurements of the pressures exerted by their molars, usually the most powerful teeth—and we can imagine how *unenthusiastic* they were during the tests.

And so is *T. rex* the all-time champ? On land, maybe. But don't go in the water. Today's great white shark has a pretty incredible bite force of 18,000 newtons, yet that doesn't come close to beating that giant croc. But its predecessor, if not its direct ancestor, the 2 million-year-old *Carcharocles megalodon*, has been credited with a maximum bite force of 180,000 newtons, ten times that of the great white, and five times that of the *T. rex*, making it the reigning champion of the bite force competition. Or is it?

Let's not forget the idea that perhaps we should not pit featherweights against heavyweights and call it a fair fight. What happens if we adjust bite forces numbers to compare creatures pound to pound? If we do that, the most powerful bite actually belongs to a tiny bird—a Galápagos finch. A recent study showed that the finch, which only weighs 33 grams (just over an ounce), has a bite force of 70 newtons. If that little finch were as big as a *T. rex*, its bite force would 320 times stronger than the dinosaur's. Let's hear it for the formidable finch!

Do goldfish memories really last only three seconds?

THE ANSWER IS NO, but the really puzzling thing is why this common belief still has a life.

Get ready to be schooled!

Goldfish are a type of Asian carp and they have been kept in aquaria and even fishbowls at least since the Jin dynasty in China, more than a thousand years ago. A later dynasty, the Tang, established the popularity of the gold version of the fish, which can also be orange, red, brown, black, yellow, or silver. While it's probably true that the vast majority of goldfish today live in captivity, there are still relatives that survive quite well in the wild.

This is the first clue that the three-second idea can't possibly be true. What would life be like for a fish with a three-second memory? Short. It finds some food, feeds, swims away, and three seconds later has forgotten not only where the food source was, what it was, or even that three seconds ago it was eating! Wherever goldfish are raised in large outdoor ponds, they have numerous predators, especially great blue herons. So a goldfish with a three-second memory might narrowly escape the plunging beak of a heron, only to swim by it again because it doesn't remember! If anything, examples like this suggest that any goldfish that did have a three-second memory would long ago have vanished from the goldfish gene pool.

Did You Know . . . The habit of keeping goldfish in bowls hearkens back to the ancient Chinese, who actually kept their fish in ponds but put them temporarily in bowls to display them for guests. But bowls are the worst possible homes for goldfish, because waste products accumulate to harmful levels and oxygen levels are usually inadequate.

If you still want to cling to the myth, you could say that maybe when they were truly wild they had enough memory to survive, but centuries of breeding in protected environments have dumbed them down, which is why those great blue herons make a very good living at goldfish farms. Domestication generally doesn't make animals smarter. Does a goldfish in a bowl really need to remember anything? You *could* make that argument, which is why it's always a good idea to see what the actual evidence is.

Let's start with the work of Roy Stokes, a student at the Australian Science and Mathematics School in Adelaide. In 2008 he started placing a piece of red Lego in a fish tank every time he fed his goldfish. It wasn't long before the fish became quite used to the Lego piece, even gathering around it within seconds before he put the food in. Then for a week he stopped putting the Lego piece in. The day he returned it, the fish gathered around in less than five seconds. So they have at least a weeklong memory.

Experiments at the University of Plymouth in the UK in the 1990s showed that goldfish would quickly learn the time of feeding and would start to press a food lever approaching that time, their activity peaking close to the exact time when a lever press would actually release food. Soon after feeding, they'd ignore the lever. Feeding was once every twenty-four hours; the fish has no trouble remembering the routine over that time span.

MythBusters on Discovery Channel also trained goldfish to remember a color pattern for more than a month. More experiments showed that goldfish trained to select certain colored tubes for food remembered the colors a year after they had last seen them. And carp (again, goldfish are a species of carp) are able to remember a fishhook a year after being caught on one, even though during that year they never saw another hook.

Is there any reason to suspect that other popular aquarium fish would be any different than goldfish? I doubt it. Experiments at MacEwan University in Edmonton, Alberta, revealed that cichlids could be trained to go to a particular area of an aquarium to be fed. The training took three days, followed by a twelve-day rest period with no further training. When the fish were reintroduced to the aquarium after the break and followed with motion tracking software, it became clear that they preferred their former feeding area.

 DON'T TRY THIS AT HOME! Swallowing live goldfish was a fad once. In 1939 a student at Harvard swallowed a goldfish on a bet. Three weeks later another student swallowed three. From there the stakes rose and more and more fish were swallowed at one sitting, reaching a hundred or more. It was a cruel game, even if it is good to include fish in your diet.

Now that your belief in the three-second memory is slipping away, I'll give you one more chance: How do we know, you might say, that they're actually remembering where or when they were fed, rather than just developing a thoughtless routine that kicks in automatically when they're faced with the same situation? In other words, they could be little flesh-and-blood robots.

I sense desperation in that argument. What is memory, anyway? If you're driving a car and face a sudden emergency, you swerve and hit the brakes at the same time, but you're not thinking about it; you do it automatically. Why? Because you've learned that's the thing to do, and it resides in your memory, but unconsciously, not consciously. Even if you want to argue that goldfish are little automatons, are they any different from you?

What animal has the biggest poop?

THAT MAY NOT BE A QUESTION that a lot of people are asking, but it does sound like a straight-forward one, doesn't it? But of course it isn't.

Are we talking about which animal produces the single biggest chunk? Or are we more interested in total volume of each dump? If a deer produces a mound of tiny pellets, do we measure by the pellet or the mound? And what about frequency—say, how much a single animal can produce in a twenty-four-hour period? And are we considering animals both extinct and living? Many angles, but in the end there has to be some sort of grand winner.

Let's consider all living and extinct animals but start with the present—the elephant. They're huge animals—African elephant males can easily weigh 6,000 kilos (more than 13,000 pounds), whereas a rhino comes in at only 2,000 kilos (a shade under 4,500 pounds). If feces scale up with increasing body size, then as far as land animals go, elephants should impress. And indeed they do.

An elephant can produce as much as 135 kilos (300 pounds) of dung every day, which is impressive but in a way not surprising, because they digest less than half of the food they eat. But of course this impressive quantity is not eliminated all at once but is shed in a series of individual dung balls. Those top out at around 16 centimeters diameter (just over 6 inches), bigger than a softball but smaller than a bowling ball.

The dung beetles got here first, I see.

Despite the incredible volume of feces that elephants produce, the land they roam is not littered with their castoffs. That's because of dung beetles. Wherever there is elephant dung, as soon as the sun sets, these remarkable insects flock in from all directions to grab a little piece of the action. Their numbers are extraordinary: in one experiment where scientists put out a half liter of dung as bait, 3,800 dung beetles were on the scene after only fifteen minutes. The beetles are going for the food value still remaining in the undigested part of the dung. Some burrow into it and eat; some dig a hole, bury it, and eat; and the most famous roll the dung into a ball and transport it away from the main mass. Some of the ball rollers are big enough to roll a tennis ball–sized piece of elephant dung.

Hard Lumps   Lumpy Turd   Cracked Sausage

Bristol Stool Chart

Elephants are herbivores. If we were to choose a carnivore as a measuring stick, one of the most impressive, though extinct, would surely be *Tyrannosaurus rex*. And, happily, a fossil *T. rex* bowel movement has been discovered and measured. Found in Saskatchewan, this particular piece of dung (called a coprolite) is 44 centimeters long (nearly 18 inches) and has a volume of nearly 3 liters, although the original was likely bigger. It's suspected that compression and drying over millions of years have compacted it. The size is impressive—it's still thought to be the biggest single piece of carnivore poop ever found—but the most striking thing about this object is that almost half of the total mass was pieces of bone. So much for the idea that *T. rex* carefully stripped the meat off its victims. It likely just gulped down huge mouthfuls, crushing the bones along with the meat.

If you're wondering why I'm not counting that two-meter-high mass of dinosaur poop in the original *Jurassic Park*? That was a movie.

 **Did You Know . . .** The Bristol Stool Scale is a handy guide that identifies seven different stool consistencies, with illustrations of each. Type 1 is separate hard lumps, like nuts, and type 7 is pretty much liquid. It was developed as a diagnostic medical tool, but it crops in scientific studies and experiments, including the debate over which animal has the biggest poop, as well as in the scientific journal *Soft Matter*.

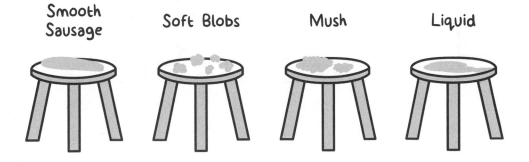

Smooth Sausage Soft Blobs Mush Liquid

If we're not focusing so much on individual pieces of dung but the total volume flushed out of an animal at one time, then both the elephant and *T. rex* are really not even in contention.

A few years ago photographer Eddie Kisfaludy snapped an incredible photo of a blue whale with a stream of reddish material behind it. The stream was about as long as the whale itself, so likely 30 meters (almost 100 feet) in length. It was impossible to judge the depth, but it was clearly a stream of blue whale feces, the reddish color caused by the algae it was feeding on. The texture of whale feces is described as fluffy or woolly, so while this bowel movement might not have tipped the scales as much as an equivalent amount of *T. rex* or elephant poop, the fact that it was 30 meters long pretty much clinches it as the biggest ever. And, like the elephant, whale poop is crucial to the local environment. Whales bring up vast amounts of nitrogen from the deep ocean to the surface and, by doing so, fertilize the ocean surface and power the food webs there.

It's apparent that judging the biggest poop is complicated by the poop's consistency, by the animal's diet, and by the likelihood of finding samples, especially in the case of extinct animals. The prize has to be given to the blue whale, but in each of these examples the coolest feature is how the poop reveals the impact each animal has or had on the other living things in its environment.

Why do zebras have stripes?

PEOPLE HAVE BEEN ASKING THIS QUESTION for a century and a half, and the final answer isn't in yet, but of the four most popular explanations, only one seems to be gaining credibility. And it might not be the one you'd think.

Let's start with idea that "stripes are camouflage." If anything, at first glance it feels like stripes makes zebras stand out. (Brown and green stripes against a mixed vegetation background would be one thing, but black-and-white?) Actually, "distraction" would be a better term than camouflage, and experiments with humans tracking and trying to catch prey on a computer screen have shown that prominent stripes make accurate tracking more difficult. But that's humans in a lab; is it different for lions and leopards? Scientists simulated the visual systems of zebras' major predators, the hyena and the lion, and found that in daylight these animals couldn't make out stripes unless they were within fifty meters (55 yards)— thirty meters (33 yards) at dusk. So it's likely a lion will have chosen a zebra by body shape or odor before it even sees the stripes.

It's also worth noting that everywhere that lions coexist with zebras, they kill more zebras than you'd expect, given the size of their respective populations. So stripes as visual protection seem unlikely.

The next theory is that stripes are signals in zebras' social life. Patterns and colors of feathers, fur, and skin *are* part of mating and social recognition in the animal kingdom. Zebras are highly organized socially, and perhaps the stripe patterns help them identify who's who. But wild horses recognize each other without the benefit of stripes; they use sight, sound, and smell, so the stripes aren't likely necessary for zebra to do the same. And non-striped zebras appear not be ostracized: a spotted zebra was seen in Zimbabwe fifty years ago and seemed to be an accepted part of the group.

Is the purpose of stripes to create air conditioning, then? This is a pretty cool idea, based on the idea that black stripes absorb much more heat than white ones. Alternating them would (theoretically) create some sort of current moving across the zebra's body. One research group asked whether stripe patterns fluctuate with climate and found that those zebras that live in the hottest areas had the darkest, boldest black stripes, and strong contrast between black and white, suggesting that the stripes might indeed be a response to the heat. But since this might also simply be an accidental correlation, a different team outfitted metal tanks of water either with striped or monochrome animal skins and measured the temperature of the water inside. There was no significant temperature difference among any of the barrels. So stripes for temperature is no longer one of the leading ideas.

And finally, might stripes discourage biting flies? You could imagine the value of reducing bites: not only would the animal be more comfortable—zebras have short-haired, fly-friendly coats—it might also help avoid fly-borne diseases. But it's hard to design clear-cut experiments to test this. The latest attempt was led by Tim Caro of the University of California, Davis, who's been after this stripe thing for years; he's even written a book about it.

He and his team went to a farm in England where both horses and zebras were being kept. The farm was a favorite haunt for horseflies, so it was an ideal setting to see if horseflies attacked zebras less than horses. Other than the stripes, the two should seem pretty similar to a fly.

Tracking the horseflies, researchers noticed that when they approached a horse, they hovered, circled a couple of times, landed, walked around, then maybe bit the horse. But when horseflies approached zebras, they did not decelerate for landing, as they did with the horses. Instead, they just came barreling in. Sometimes they'd swerve away at the last moment, or they'd touch the hide briefly before moving on, but sometimes they flew directly into the zebra's body headfirst, bounced off, and flew away. In the final tally, the horseflies actually touched zebra skin more often than horse but landed much more often on horses.

You're not fooling anyone.

To see if this difference had something to do with a zebra's stripes, the team clothed horses in striped coats. These horses received many fewer bites on their bodies but the same number on their uncovered heads. So it wasn't the whole animal the flies were rejecting, just the striped part.

One suggestion is that stripes, because they're unusual in the fly's environment, might disrupt the way the fly's eyes and brain work together. Stripes might make the fly see movement where there isn't any, or vice versa, causing the fly to abort its landing—or just smash into the zebra, as they seemed to do. The closest human example is likely the rotating barber pole, which gives you the feeling that the pole itself is moving up or down, depending on which way it's turning.

And don't forget that a zebra's stripes are different widths and contrasts, in some places horizontal on the body, in others places vertical. All that inconsistency might worsen the situation for the fly.

Tim Caro has also mused about the possibility that the stripes create another sort of illusion: they may look like a set of parallel dark objects with spaces in between that flies can fly through—or not, as it turns out.

If this *is* true, why haven't more animals that are susceptible to fly bites evolved stripes? Even the mammals that share the African environment with zebras haven't done that. And what about the okapi, that relative of the giraffe, which has stripes on its hindquarters only?

Charles Darwin himself was involved in the first public scientific disagreement over the zebra's stripes—150 years ago. We're still at it today, but maybe for the first time we have a convincing scenario. We'll see.

History Mystery

What was the Black Death?

THE BLACK DEATH, also known as bubonic plague or the Black Plague, was likely the most deadly disease to strike humans in the last two thousand years. From 1347 to 1352, it wiped out about 20 million people, half the total population of Europe. For comparison, the influenza pandemic (the "Spanish Flu") that struck at the end of World War I, killed somewhere

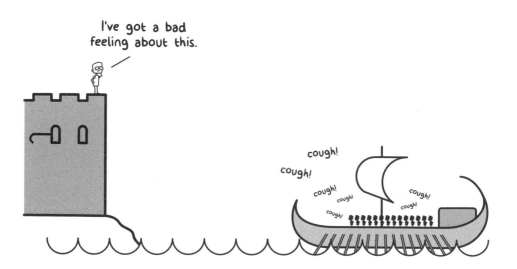

between 50 million and 100 million, but that amounted to only 5 percent of the total world population.

In fact, the disease wasn't limited to those five terrible years: it moved around Europe and the Middle East for centuries after, flaring up here and there. But we know a lot about the 1347 European version.

It began in the fall when a fleet of galleys sailed from the Black Sea and docked in Sicily. Sicilians were shocked to find the sailors onboard either dead or dying, covered with black, pus-filled boils. Harbormasters ordered the ships off the docks, but it was too late. Once established in Sicily, the disease began to spread throughout Europe over the next four years.

Once a person caught this devastating bacterial infection, their chances of survival were slim: lymph nodes in the armpits, neck, and groin swelled and blackened, and the patient was overwhelmed both by the bacteria and the body's own immune response. The numbers of dead are shocking enough, but descriptions from the time tell the story more vividly: ships drifting helplessly on the open sea, their entire crews dead; mass graves with bodies stacked five deep, as one writer put it, like layers of lasagna; villages practically deserted; doctors helpless in the face of a disease they couldn't begin to understand. The Black Death wound down in 1352, but its impact was felt for centuries after.

Science Fact! *People do indeed still die from the bubonic plague. In May 2019 a couple in Mongolia died of the plague after eating raw meat and kidney from a marmot, an animal known to harbor Yersinia pestis. This is such an odd set of circumstances that there seems to be no real risk of a serious outbreak as a result.*

The mechanics of the disease, at least as we reconstruct it today, were straight-forward. Rats were abundant on ships, towns, everywhere there were people. Rats had fleas. If those fleas were infected with the bacterium called *Yersinia pestis*, the rat would become infected, too, and would soon die. But before it did, other fleas would ingest blood (and bacteria) from it, then move on to other rats to repeat the process. As the rat population dwindled, the fleas switched to humans. Once established in a human, the bacteria usually had reached a dead end, but occasionally could be transmitted from person to person by breath. This was the so-called pneu-monic version of the disease, even deadlier than the flea-based version. Humans in Europe had no prior immunity to the bacterium, so the plague ran roughshod over Europe: Sicily in late 1347; France, Spain, Portugal, and England by June

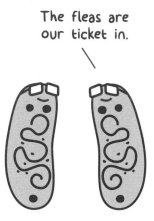

The fleas are our ticket in.

1348; Germany, Scotland, and Scandinavia from 1348 to 1350—a tidal wave of disease at a time when horse-drawn vehicles were the fastest things on land and sailing ships on the sea. This is one reason some skepticism has arisen: How could a disease that requires so many time-consuming steps achieve this lightning overland spread?

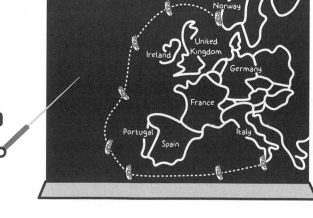

Spread of the Black Death

While some researchers have suggested that the Black Death might have been a wholly different disease than we think, genetic testing of teeth from graves all over Europe has found traces of the *Yersinia pestis* bacterium—although there might have been two or even three types, and the routes they traced across Europe might have been different from what we have assumed.

Did You Know . . . The Black Death likely provided the first ever opportunity for biological warfare. Before the disease reached Europe, the Tartar army, laying siege to the city of Kaffa (now Feodosia), in the Crimea, catapulted bodies of those killed by the plague over the city walls to encourage the city dwellers to surrender. It definitely wouldn't have been fun to be inside the walls, but some modern experts doubt the tactic would have been as devastating as it sounds.

Let's dissect the mystery. What about the rat? Black rats were common in the 1300s, but it is very difficult to fit them into the picture in every country that was ravaged by plague.

For instance, black rats are not cold-weather animals. They flourished in southern Europe, but Italy's climate is very different from Norway's, and the Black Death hit Norway hard. Norwegian archaeologists have found what are likely black rat bones in the medieval remains of coastal towns, likely from rats that arrived there on ships, but they've found few bones in rural areas or inland towns, suggesting rats couldn't have formed a network continuous enough to support the spread of the Black Death.

We also know from more recent but less serious episodes of plague—like one that spread outward from Hong Kong in the 1890s—that rats are seen in great numbers as they become infected and die. Chroniclers of the disease describe rats falling from roofs and piling up in the streets. But there are no such descriptions in the 1350s from Norway or even farther south, in England. But if rats weren't involved in the Black Death, what was?

In the last ten years or so researchers have begun to suspect that human parasites, like the *human* flea, not the rat flea, and the human body louse, might have played a key role. In medieval times, clothing and bedding were likely full of both of these. As people traveled, they'd take at least their clothing, and likely their bedding, with them. This could explain the rapidity of spread: it doesn't require the local rat population to be completely infected before the disease spreads to humans. It also would explain why one household could be struck by the plague but not the one next door. (Rats' territories are large enough to include several houses at a time.) Researchers have recently developed a mathematical model of the spread of the Black Death and several subsequent plagues using human parasites rather than the rat fleas and found that the model mimicked the actual spread of the Black Death much better.

Why is there still so much interest in a plague that happened more than six hundred years ago? Because the plague bacterium, *Yersinia pestis*, still exists, causing small outbreaks around the world. It is still capable of killing humans. The better we understand the bacterium and the way it can spread from human to human, the better prepared we will be in the event that climate, parasites, and antibiotic resistance line up to allow another serious outbreak.

Science <u>Fiction!</u> It's commonly believed that the rhyme "Ring Around the Rosie" was inspired by the Black Death. The lyrics "ring around the rosie" is supposed to describe the spots on the skin, "a pocket full of posie" refers to the use of flowers to ward off the disease, "husha, husha" (or "ashes, ashes") imitates the sound of sneezing, and, of course, "we all fall down" means death.

But the poem first appeared in print in 1881, and it wasn't linked to the Black Death until 1961. If the rhyme was really inspired by the Black Death, then you'd have to believe that children secretly rehearsed it for more than five hundred years. An unlikely thought.

Part 3
Peculiar Phenomena

Does time slow down when we're in an emergency?

NEARLY FIFTY YEARS AGO, two psychiatrists, Russell Noyes and Roy Kletti, interviewed people who had survived near-fatal incidents. These included falls while mountain climbing, near drownings, car accidents, and battlefield explosions—incidents that arose instantly in which death seemed inevitable. When the psychiatrists asked the people to describe how they had felt in the moment, two things stood out.

Most of them felt as if time had stood still, or at least slowed dramatically. A stock car driver recounted turning over and over in the air ten meters up: "I remember thinking that death or injury was coming but after that I didn't feel much at all. It seemed like the whole thing took forever. Everything was in slow motion . . ." A young driver lost control of her car: "During all of this, time stood still. It seemed to take forever for everything to happen . . ."

The other thing that many noted was that, during the life-threatening events, thoughts raced through their minds at incredible speed, much faster than normal. These included memories about their lives as well as strategies to cope with their situations—all mingled together in a few seconds.

There is no reason not to trust these accounts of the experience, and Kletti and Noyes were by no means the first to report this, but it is hard to understand how either could happen, especially how time could seem to slow down. After all, time flows at a consistent rate, independent of us; our phones or watches tell us that. But we're not talking here about time as a feature of the universe, but time as we experience it.

Did You Know . . . Some animals are very different from us in their ability to "slow things down" visually. When we watch video, our mind convinces us that what we're seeing is a smooth flow of images, not a series of stills. But a dog sees that video as a steady flicker, and some animals are much more acute than that. Flies likely see our movements as hopelessly awkward and slow. They're like Neo dodging bullets in *The Matrix*.

Of course, these two impressions, the slowing of time and the quickening of thought, are completely subjective. That makes them very difficult to study scientifically. However, one experiment by neuroscientists about ten years ago tried to evaluate both claims. The idea was to subject volunteers to what would feel like a life-threatening situation while testing to determine if their minds were racing. They would then be asked how long they thought the event had taken.

The test of mental speed was a device strapped to the wrist that displayed either red numbers on a neutral background or the reverse, a red background with neutral numbers. If these two displays alternate rapidly enough, it becomes impossible to distinguish the digits from the background. For most of the people in this experiment, that happened when they alternated faster than every thirty- to fifty-thousandths of a second. In preparation for their emergency experience, each person's display was set about six-thousandths of a second faster than their threshold, meaning that normally they wouldn't be able to identify the numbers. But if their minds were working faster, they would.

The "life-threatening" experience (which was actually safe) was the suspended catch air device (SCAD) diving tower at the Zero Gravity Thrill Amusement Park in Dallas, Texas. On this ride, you free-fall 31 meters (more than 100 feet) into a safety net. The drop takes about 2.5 seconds. If minds ran faster during the fearful moments of the drop, the volunteers should have been able to distinguish the digits on their wrist displays, even though they were displaying

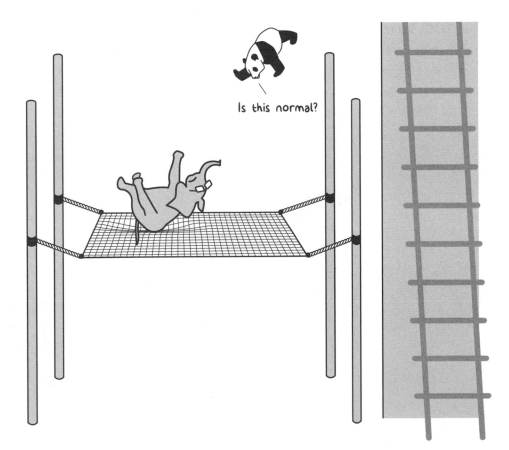

Is this normal?

above their threshold. But none of the participants could, suggesting that, at least in this case, brains worked at normal speed.

When asked to compare their own free falls to others that they had watched earlier, however, the participants felt that theirs had taken significantly longer. So it looked as if their retroactive impression was that time had slowed, despite the fact that during the fall their minds were working at normal speed. This led the investigators to argue that claims of time slowing during emergencies were actually the result of inaccurate memories, not of real-time experiences.

As ingenious as this experiment was, it clearly doesn't exactly mimic an actual near-death experience. (Subjecting volunteers to a real emergency would be unethical). The diving tower didn't catch people unaware, as true emergencies do; as well as assuming that thousands of test subjects had done it already, the volunteers had watched others do the same thing minutes earlier.

Did You Know . . . Many studies have shown that our perception of time doesn't always match how much time has actually passed. The same interval can feel different if you're concentrating or not, if you have a fever, or as you age. Accepting that the time we experience may not track with clock time is an important first step to believing that time might seem to slow down at extraordinary moments.

I'm on island time!

So, of the two ideas, time slowing and thought speeding up, it might be easier to explain the latter. We know that in *real* crises, hormone responses connected to the "flight-or-fight" syndrome can accelerate our ability to react, which makes it possible that people who have claimed their thoughts accelerated, too, are reporting exactly what happened. But we'll have to wait for the definitive experiments that will either confirm or disprove that.

Is there a way to choose the fastest checkout line at the grocery store?

You're at the supermarket and, as usual, you're in a hurry. When you get to the cashiers, there are lines at every one, some longer, some shorter. A quick glance and you choose the one

that looks the fastest. Two minutes later you're cursing your choice and glancing around to see if it's worth switching to another line. There are at least two things making you miserable: one is the challenging mathematics of lineups and the other is your own impatience. The good news is that you—or the supermarket—can do something about both.

First, the math. How did you choose what you thought was the fastest line? You likely knew that it's not enough simply to count the customers ahead of you, because you can't be guaranteed that each cart is going to be checked out at same speed. You might have estimated the number of items in each cart, then compared the number of items—rather than the number of carts—in each line. That's smart, but the math is even trickier than that.

American mathematician Dan Meyer spent time in a supermarket to identify all the variables. In the Whole Foods Market he surveyed, the average time for an interaction between the cashier and the customer was more than 40 seconds, but each item took only 3 seconds to process. Those numbers explain why two carts stacked to the brim with items may look like a roadblock but might be faster than three or four half-full carts. One cash machine's checkout data revealed that, yes, you can process nearly fourteen food items in the time it takes a customer to pay a bill. This doesn't bode well for the express lane, which features many people with a small number of items, exactly the ratio to guarantee slow movement.

You'd think the choice of cash or credit (or debit) could easily affect the speed of the transaction, and indeed Meyer found that, except that he claimed that cash was faster. He argued that some people were absolutely befuddled by their credit cards, much more than any customer with cash was.

Meyer's analysis of the supermarket line was originally published in 2009. Since then slicker processing of credit cards, especially with smartphones, would suggest that the cash-to-credit ratio might have changed by now. The advent of self-checkout might also have changed the picture, but while that option might seem that you're saving time, it's questionable whether you can check yourself out faster than a cashier. Of course, from the store's point of view, self-checkout is beautiful: fewer cashiers, same revenue.

Did You Know . . . One study suggested that Americans spend about half an hour a day waiting.

One way stores can lift the math burden from their customers is by using one long or "serpentine" line. Even better is the single serpentine line feeding several cashiers, which both eliminates the stressful choice and speeds up the checkout process. But the experience of lining up, the experience of *waiting*, can still be frustrating.

Psychologists Dan Kahneman and Ziv Carmon illustrated just how agonizing the supermarket-line experience can be. Their analysis showed that a customer in line begins at a low point, having just joined the line with no real idea of how fast it was going to be. Then a series of ups and downs follows, with every forward movement of the line triggering a short burst of elation, followed almost im1

mediately by a relapse into frustration as movement stops. The relapse occurs much earlier than any reasonable expectation of the next move forward. All this back-and-forth is exacerbated by the ongoing concern that lines to the left and right are moving faster. But as the line edges closer to the cashier, mood gradually lightens, until finally you're there!

I've checked out.

Did You Know . . . The theory of lining up—"queueing theory"—was invented by a Danish engineer, Agner Krarup Erlang, in the early 1900s. This wasn't just before smartphones; this was practically before phones of any kind. In those days phone calls went through a central switchboard. Erlang tackled the challenge of deciding how many phone lines the city of Copenhagen needed to be able to handle every phone call. He had two goals: Don't make anyone wait too long for their call to be connected, but don't hire surplus telephone operators. The mathematics created to solve problems like that are part of queueing theory.

As emotionally turbulent as the entire experience is, Kahneman and Carmon showed that the final arrival at the till, and the overwhelming joy that accompanies it, is what colors people's memories of lining up the most. But there are clever ways of changing your experience of the wait, of reducing your anxiety and frustration long before you check the checkout.

One way is to entertain you: amusement parks like Disneyworld do this extremely well, making the space where you're waiting as entertaining as most actual exhibits in other parks. It's much harder for grocery stores to provide this sort of entertainment, although other kinds of retailers sometimes use serpentine or multiline dividers stuffed with small items that will distract you (and perhaps fill your cart) as you move past. But you can also entertain yourself with things like a smartphone or a magazine. Any sort of distraction alters your perception of the amount of time you spend in the line. In the absence of entertainment, a precise estimate of when you'll be served makes life in line much easier than having no idea when you'll get to the front. But while that works for subway trains (sometimes), it can't be done in the supermarket.

I'm going to trip him up.

There are no easy answers, but it's worth remembering that once you get to the front of the line, you will feel much better. As long as the person ahead of you doesn't have any coupons, that is.

Why do headphone cords always get tangled?

Knots are crucial for tying shoes but totally frustrating when you find them in your headphone cords. It's painstaking drudgery to untie them but fascinating to consider how those knots got there in the first place, especially since your headphones were just sitting quietly in your backpack.

About twelve years ago two physicists at the University of California, San Diego, Dorian M. Raymer and Douglas E. Smith, did just that, publishing a scientific account of knot formation called "Spontaneous Knotting of an Agitated String." Here's how they designed their experiment.

First they built a transparent box, 30 centimeters (almost 1 foot) on each side, with rotating metal rods attached to two sides that caused the box to spin.

I'm in a real bind

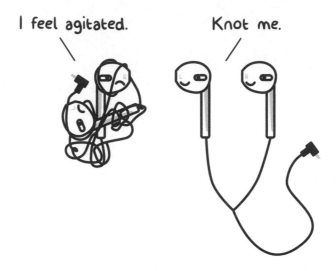

I feel agitated.

Knot me.

Then they simply dropped a piece of string in, closed the lid, and spun the box. The effect was like clothes being tossed in a dryer.

For knots to form, the free ends of the piece of string must cross over the rest of it at least once. Raymer and Smith found that the length of the string and the speed of rotation were both critical factors. The string itself was about a tenth of an inch in diameter. If it was around 46 centimeters (18 inches) or shorter, it wouldn't form knots, probably because the free ends just didn't encounter the rest of the string that much. But longer strings formed knots beautifully: if they were 1.5 meters (5 feet) long, knots formed 50 percent of the time.

The speed of rotation didn't seem to matter much until it reached three rotations per second. At this high speed, the string spent most of its time pressed up against the sides of the box, reducing its actual tumbling time dramatically. The same happened when the size of the box was reduced by a third: there wasn't enough room for the string to tangle itself.

Raymer and Smith also found that the flexibility of the string mattered. A stiffer string generated many fewer knots, because, again, as with the faster rotation, the string found itself wedged against the walls of the box. As you might expect, more flexible strings were knottier because they flipped and flopped all over themselves. After 3,415 trials(!), analysis of the data suggested that segments of string tend to lie parallel to each, and either end of the string manages to weave itself over and under the segments. It's not simple; even this basic mechanism produced 120 different varieties of knots. And, once knotted, physics makes it clear that the knot is unlikely to be undone easily. After all, the string was originally free to flop around but now is wound around itself; extracting the string from the knot would require a significant amount of energy. At the same time, more knotting is happening.

Science _Fact!_ *Umbilical cords become knotted in the womb only about 1 percent of the time. Raymer and Smith speculate that the percentage is that low because there just isn't room in the uterus for the cord to become tangled.*

And so what does this say about those headphones in a messy heap in the bottom of your bag, bearing in mind that, because they're Y shaped, they were even more likely to become knotted? You could choose new ones with thicker cords, or shorter cords, although of course convenience would take a major hit in either case. You could wear them around your neck. You could even wrap them around a stick. You could buy the cordless versions.

Or you could just leave things the way they are and marvel at the complexity that can be created by just a few hours in a backpack.

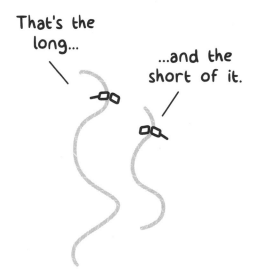

That's the long...

...and the short of it.

Did You Know . . . Legend has it that the most difficult knot ever made was used to tie a cart to a post by a man named Gordius, thousands of years ago in Phrygia. The Gordian knot was so elaborate, so impossible to untie, that an oracle predicted anyone who could untie it would become the ruler of the world. Alexander the Great arrived in the town, was told the story, and decided then and there to be the man who solved the knotty problem. He failed to do it by pulling the loops apart and twisting the rope, so finally he simply took his sword and cut it apart, arguing that no one had said exactly how it should be untied. (You could probably use the same technique to untie the tangle in your headphone cords, but only once.) It was kind of a cheat, but who's going to argue with Alexander the Great? He did go on to rule the world after all.

Hi Gordie.

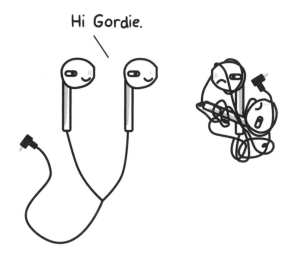

Can wasps turn cockroaches into zombies?

I MUST HAVE WATCHED TOO MANY EPISODES of *The Walking Dead*. Zombies mean only one thing to me: staggering, wild-eyed, slavering former humans mouthing guttural moans in the never-ending search for fresh human flesh. And, thankfully, they're pure fiction. Well, at least the human zombies are. But in the insect world, this hideous concept is much closer to reality.

I'm dead tired.

Zombies, by general definition, are beings that return from the dead but have no will. The "no will" part of that is crucial when it comes to insect zombies, of which there are a number. One is the result of the relationship of the emerald jewel wasp and the cockroach. If you're a roach, this is a horrifying story; if you're a scientist, it's a beautiful example of an evolutionary arms race, a race between predator and prey.

The wasp is much smaller than the roach. It's 2.2 centimeters (a little less than an inch) long, while the roach is nearly double that length and several times wider. But the size difference is irrelevant: the wasp has the weapons.

117

First strike: the wasp lands, grabs the roach just behind the head, and injects venom into the thorax, the insect's middle segment. The roach's front legs are immediately paralyzed and it slumps forward, resting its head on the ground. The roach will recover, but this short-lived inability to move dooms it, because it gives the wasp time for the crucial and precise second strike.

Second strike: the wasp now moves forward to the roach's head and searches for the exact spot to inject a second shot of venom. This is directed with surgical precision into a central set of neurons in the roach's brain, and the result is both weird and deadly. First, the roach starts a prolonged bout of grooming, threading both its front legs and antennae through its mouth, over and over. This is normal behavior—it is especially important to keep the antennae clean and able to respond to the roach's environment—but in this case it goes on for an unbelievable thirty minutes. Long enough for the wasp (a female) to dig a burrow.

When she returns to the roach—which has made no attempt to escape in her absence—she bites the roach's antennae in half and drinks the hemolymph (the insect equivalent of blood) that leaks out of the severed ends. (For nutrition? No one seems to know.) Then she grasps one stump of an antenna and leads the roach, like a dog on a leash, into the burrow. The roach is still perfectly able to walk and will not resist being led but cannot initiate walking on its own. It is a zombie. And unfortunately things are about to get worse.

Walking the Dead.

Science Fact! *Cockroaches have survived for hundreds of millions of years because they have amazing survival abilities. For example, if you cut off a roach's head, it can still live for weeks. Not so with humans. We can't breathe if our brains are separated from our bodies. We also have a high-pressure system of blood vessels, and so, if we lose our heads, we bleed to death. But cockroaches can withstand both these risks because they breathe through holes scattered all over their bodies and their vascular system is low-pressure, which means their blood can clot quickly so they don't hemorrhage to death. Not only that, the nervous system in their bodies allows a headless roach to walk around, scratch, and even learn to avoid an electric shock for at least for a couple of weeks. That's why they have a reputation for being indestructible.*

Once the wasp and roach arrive at the bottom of the burrow, the wasp lays a single egg, close to the roach's thorax, on the upper joint of its middle leg. The wasp then leaves the burrow and covers the entrance with pebbles. Her work is done.

When the egg hatches a couple of days later, the larva immediately bites a hole in the roach and starts to drink the hemolymph. After a few days of that, it burrows into the still-living roach's body and proceeds to consume its internal organs. And if you thought this was already grisly enough, the wasp larva consumes those organs in a sequence that ensures that the roach will live as long as possible. The wasp-to-be finally spins a cocoon inside the now hollowed-out body, pupates, and a month later emerges. The now-adult wasp makes it way out of the burrow to mate, find another roach, and continue the process. It will live long enough to attack several more roaches. As has been said, every emerald jewel wasp is "of cockroach born."

But as grim as this picture is, the roach isn't *completely* helpless. While the wasp uses chemical warfare, the roach responds with old-fashioned martial arts tactics: it fights back. Sensing the wasp's presence, the roach "stilt stands," straightening its legs and raising itself off the ground. From this stance it can use its antenna to track the wasp's position and turn away from it. If it feels the threatening touch of the wasp's antennae as it prepares to sting, the roach will bring its hind leg forward and then, in a sweeping strike, try to knock the wasp away. It's a violent baseball swing of the leg and, if accurate, sends the wasp flying.

But leg swinging isn't the roach's only defense. In close quarters, it will rake the wasp's body with the spines on its legs or even try to bite the wasp. Putting up a stiff defense sometimes seems to discourage the wasp from continuing the attack, and in one lab study nearly two-thirds of the cockroaches that defended themselves succeeded in escaping—not a bad percentage.

If this is indeed an evolutionary arms race, then who's really ahead? The wasp has clearly evolved precise techniques to subdue the roach. While relatively effective, the roach's defenses seem kind of old-school in comparison—and they may simply be the same tactic they developed to fend off all predators. They have not, like some types of prey in other insect battles, evolved defenses against toxic venoms.

So, for now, I'd give the edge to the wasp. But as we know, cockroaches always find a way to persist.

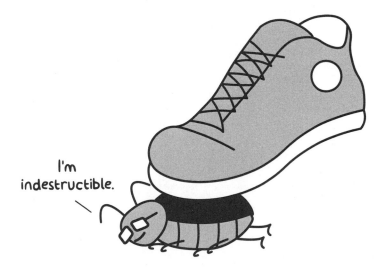

What is photographic memory?

MOSTLY, WHEN PEOPLE TALK ABOUT PHOTOGRAPHIC MEMORY, they mean the ability to store in our memory, with photographic-like precision, images that remain stable and can be scanned, even a long time after the event. That's a very high standard, and as far as anyone knows, there's no undisputed evidence there is a human who can do that.

The closest we've come was in an account published in 1970 by a Harvard psychologist named Charles Stromeyer III. He did tests with a young artist, Elizabeth, who claimed to be able to look at a painting for a few seconds, then turn to her empty easel and "project" the painting onto it, in every detail.

Stromeyer's experiments with Elizabeth were subtle and clever. He worked with what are called "random dot stereograms." These look like extremely complicated QR codes, the images you can read with your smartphone's camera. But his images were much more intricate: 10,000 dots each. They were designed to be paired in a stereo viewer,

Lest we forget!

one seen by the right eye, the other by the left, and an image—in this case a letter of the alphabet—would emerge from the combination of the dots. For this to work, you would have to be looking at the two images simultaneously.

But not Elizabeth. Stromeyer first introduced a ten-second delay between the stereograms, then a ten-minute delay, and both times Elizabeth instantly identified the hidden letter. The only way she could have done this would be if she had kept the original image perfectly in her mind, then superimposed it on the second image. And that meant she had a photographic memory of the first image—for as long as ten minutes. Elizabeth still had no trouble when Stromeyer repeated the experiment with a much longer twenty-four-hour delay between the right and left images.

At one point in describing this work, Stromeyer admitted, "Obviously more research is needed." But then he married Elizabeth, and that was the end of the experiments! Today confidence in this work has waned because we heard nothing more about Elizabeth, and there have been no people like her found in the fifty years since.

But there are other intriguing examples of unusual people with unusual abilities. Their memories are amazing if not quite photographic.

Some have what's called "eidetic imagery," sometimes referred to as "eidetic memory." This phenomenon happens mostly with young children, maybe 10 percent of them. Kids who have eidetic imagery can see a picture once, then can scan it in their minds as if it were still in front of them while answering questions about it. Eidetic imagery is fascinating, but unlike our concept of photographic memory, errors creep in, and it lasts only a few minutes. Even a simple blink of the eyes can destroy the eidetic image forever.

This makes eidetic imagery very different from standard memory. With the latter you might have a picture in your head of a familiar room, like your bedroom, but there are likely details you won't recall if asked to describe the room. But this memory, despite its flaws, lasts—instead of vanishing like an eidetic image.

It's puzzling that children with eidetic memory usually lose the ability as they become adults. This may have something to do with how the image, or the memory, is coded. Children who

were asked to scan pictures and also name the features they were looking at had great trouble forming an eidetic image; language interfered with what had been purely a visual process. Possibly adults have even more interfering processes going on, making it impossible to create such mental images.

Then there are individuals with some extraordinary mental abilities but also several intellectual challenges. They're known as "savants." Most savants have prodigious memories. Kim Peek, the inspiration for the character Raymond Babbitt in the movie *Rain Man*, was one of these. Peek was born with several significant brain abnormalities but had an unbelievable ability to memorize. He could read a book in about an hour, scanning the left page with his left eye and the right with his right. And he could retain vast amounts of what he had read; some estimates suggest he could remember something like 12,000 books. If you gave him your birth date, he could tell instantly what day of the week it was. He knew the details of hundreds of classical music compositions, including when and where each was composed and first performed. Kim Peek's broad range of knowledge and abilities set him apart from most savants, who are usually restricted to a single specialty, like knowing the days of the week for any date.

 **Did You Know . . .** One of the most bizarre feats of memory ever recorded was the claim that a group called the Shas Pollak, a group of Jewish memory experts, knew the exact location of every word in the 5,000 pages of the Babylonian Talmud, the book of Jewish religious law. If you stuck a pin into the book penetrating as far as, say, page 245, a Shas Pollak individual could tell which word the pin was pointing to. Sadly, as fascinating as this is, the only reports are second-person, word-of-mouth accounts.

Memory "champions" are another group who have unusual abilities for recall. They can remember the order of cards in a deck after going through them only once, or enormous sequences of numbers, like 550 digits, in five minutes. But it's their sophisticated methods and training that are extraordinary, not their actual memories. They develop systems for sorting cards or numbers into groups, then collate those groups into larger sets until they have an organized bank of data that they can retrieve easily. But, again, it's not a photographic memory.

The idea of a photographic memory, however, hangs on. It's attractive to imagine having a splendid memory that would allow you to capture an image and keep it in your mind to explore over and over again. But this is definitely a case of being careful what you wish for. Some people with extraordinary memories complain that they can't escape them: remembering every set of numbers you have ever seen or every experience you've had, both good and bad? Torture!

Why can't we break spaghetti in half?

PERHAPS YOU'VE NEVER WONDERED ABOUT THIS. Perhaps you think you actually *can* break a piece of spaghetti neatly in half. If so, try this. Hold a piece of dry spaghetti by the ends ("by the ends" is the crucial part: you can't have both hands in the middle) and slowly bend it until it snaps. It shattered into three or more pieces, didn't it?

You can do that again and again, and you will get the same results. But why?

That is a question that occupied one of the greatest scientific minds of our era—for several hours, at least. Nobel Prize–winning American physicist Richard Feynman spent the better part of one evening snapping dry lengths of spaghetti in an effort to understand why they never broke into just two pieces, the way a pencil would, but into three, four, or even more pieces. As great a scientist as Feynman was, he didn't figure out what was going on that night.

Give me a break!

Feynman died in 1988, but in the last few years scientists have figured out the answer to that question. As you slowly bend the spaghetti strand, it experiences more and more stress until it breaks. But when it snaps, the stress is suddenly released, and the two ends where the break occurred snap back so fast that a wave travels along the strand. That wave distorts the strand so much that it breaks again (or maybe even twice more), and you have bits of spaghetti all over the floor. It doesn't matter whether you break it fast or slow, the result is the same.

You *can* break the spaghetti in two if you grip the strand with your hands together in the middle, because your hands will then absorb that wave action. But there's only one way that you can break the spaghetti into only two pieces if you are holding it by the ends. However, you will likely need some special lab equipment to do it.

I'm stressed.

That's better.

Let's do the twist!

Two young scientists at MIT, Ron Heisser and Vishal Patil, built a machine that would *twist* the spaghetti before bending it. They found that if, before it was bent, the spaghetti was twisted enough—about 270 degrees, or three-quarters of the way to a full turn—it would snap cleanly into two. The explanation? Twisting the spaghetti puts it under stress the same way bending it does, but a different kind of stress. When it's bent and finally snaps, some of the energy that normally would cause additional breaks is instead used up to untwist the strand. Once that happens, there isn't enough leftover energy to cause any further breaks. They also noted that a twisted piece of spaghetti breaks with less bending than an untwisted strand.

All this is true of a single piece of spaghetti, but what happens if you hold a fistful and snap all of them simultaneously? You might expect that the wave energy generated by snapping each one would be dissipated or absorbed by the neighboring strands that it's in contact with, and that therefore with less energy available to whip the free ends around, no extra pieces would break off. (And each strand on the inside of the pack would have less room to snap back, further reducing breakage of the middle pieces.) You might expect that, but in my lab, which happens to be the kitchen table, a handful of sixty-seven full-length pieces of spaghetti shattered into two handfuls plus *ninety-three* extra pieces, suggesting that the same sort of thing happens even with a crowd of spaghetti.

My kitchen experiments also proved that no matter what the pasta—fettucine, linguine, vermicelli, spaghetto quadro, capelli d'angelo, and even bucatini (the one with the hole running down the middle)—all shatter into three or four pieces when bent and broken.

Yes, I'll have the *pezzi di pasta*, please!

Delizioso!

Why does a return trip always feel shorter?

ANY TIME YOU TAKE A TRIP to an unfamiliar destination and return the same day, you're likely to experience what has been called the "return trip effect"—that is, it seems to take much less time to return home than it did to get to the destination in the first place. It doesn't appear to

matter how you travel: by car, by bicycle, or by foot; nor is the length of the trip crucial, although it should be at least twenty minutes or more to work best. This "return trip effect" is one of the most reliable and striking examples of how our sense of time can be warped.

There are a number of possible reasons why this happens. What if you have to be at the destination at a particular time? Then every minute that passes will seem like an eternity, but coming back, in the absence of that intense focus on the passage of time, the trip seems quick and uneventful. "Uneventful" might be the key word: our *judgment* of time past is influenced by how many events *seem* to have been packed into that time.

Even if we aren't in a hurry, a trip along an unfamiliar route is full of novelty: a tree by the side of a road, a ramshackle house, a sharp curve in the road. Coming back, even though you're experiencing all these from a different perspective, you're familiar with them and need not record them again. We aren't likely to remember so many "events." Indeed, the great American psychologist William James once pointed out that judging time as it's happening demands you pay attention, but judging time that has already passed depends on memory. The two are very different.

There have been experiments done to determine exactly what's going on in the return trip effect. In one, Japanese researchers had a cameraman shoot a set of twenty-six-minute videos as he walked. The first recorded walking a distance of 1.7 kilometers (a little over a mile) through city streets. The second was the exact return trip, taking the same time. The third trip followed a different route back, but one that was the same distance and took the same time.

Volunteers who watched the movies were asked two questions. First, as they watched, they had to estimate when the first three minutes had passed, then the next three, and the next. Then, at the end of both the outward- and inward-bound trips, they had to estimate which had seemed longer. Remember, some of them had watched the filmmakers retrace their steps exactly, but others had watched them return by a different route.

The results would not have surprised William James: both groups were pretty much the same at estimating the passage of three-minute chunks of time, but those who retraced their steps estimated the return trip to be much shorter; those who took a different route home didn't. So people's accuracy at keeping time as it passed seemed to be about the same, but remembering which leg of the trip was longer was different. The fact that those who took a different route back experienced new things on both legs of the trip, while those who retraced their steps didn't, suggests that familiarity with the route or the lack thereof can play games with your memory of a trip's duration. This might also explain why the return trip effect seems not to happen with a commute. After a while, it's all familiar, and your attention is occupied by other things.

While this all seems to make a certain sense, a different study came up with a different conclusion.

This one, in the Netherlands, was set up to investigate whether our expectations, not our memory, was the crucial feature. They ran three experiments. In one, people who had traveled to a housekeeping fair (possibly because free stuff was being given away) were asked how long the return trip had taken compared to the trip out. A majority felt the return trip was shorter, but it seemed to have nothing to do with familiarity. The number of landmarks they recognized from the outgoing trip had no effect on the estimated length of time of the return, but—and here's the key point—most thought the outgoing trip took longer than they had *expected*.

This test seemed to indicate that if your outgoing trip seemed longer than you expected, that leads you to ramp up your expectation of the time needed for the return trip. Maybe you overdo it, and the return then seems shorter than you thought it would be.

A second version involved bike riders traveling to and from a camp in the forest. The group was ninety-three first-year university students (although eight had to be dropped because they got lost on the return trip!). As in the Japanese experiments, the group was split into those who took the same route back and those who took a different one. This time, there was no difference between the two groups. If familiarity with the route is the cause of underestimating the return trip, then those who backtracked should have judged their return to be shorter than those who took a different road. But they didn't. However, again, their feeling that the outgoing trip was longer than they had expected made them anticipate the return trip to be longer as well, even though it turned out not to be.

A third version seemed to confirm the idea that expectations are key. This time students followed a bike ride shot from a camera, but as part of the study, prior to watching the video, students read an account apparently written by a student who had already watched it that said, "Phewwww, that video took a lot longer than I expected."

Students who read this showed no return trip effect, apparently because they were prepared that the outgoing trip was going to seem long, so it wasn't longer than they had expected. Having no need to adjust expectations upward for the return trip, it didn't seem shorter.

Despite the Dutch experiments, it seems likely that familiarity, as well as expectations, might contribute to the return trip effect, depending on the circumstances—and there might even be other, as-yet-unidentified factors. If nothing else, these experiments demonstrate how the passing of time is a very personal thing.

History Mystery

Where did the Easter Island stone heads come from?

EASTER ISLAND is the most isolated piece of habitable land in the world, 3,700 kilometers from the coast of South America. And it's home to a set of one of the world's most impressive ancient monuments. The stone heads, called *moai*, are instantly recognizable with their overhanging foreheads, long, sloping noses, and prominent chins. (They actually have bodies, too, many of which are now partially buried). There are more than nine hundred of them,

many unfinished. As tall as 10 meters (about 33 feet), and weighing up to 82 tons, they stand in groups, expressionless. Some of the unfinished and apparently abandoned statues are even bigger.

Archaeological evidence suggests that Polynesians, who were of course incredibly skilled seagoing people, likely arrived on the island sometime before the year 1000, although that date is questioned. Most archaeologists estimate that the statues were built later, likely between 1200 and 1500.

Like the English monument Stonehenge, one of the most remarkable features of the stones is that they weren't simply carved and then erected on the spot where we find them today. They were moved from various places on the island, which given their weight and the distances involved—up to 10 kilometers (more than 6 miles)—must have been a huge challenge for a society lacking the wheel.

Two theories have been advanced to explain how the statues were moved: either lying down and rolled into place on logs, then levered upright, or, strange as it might sound, walked into place vertically. The idea of walking upright stones was suggested by archaeologists Terry Hunt and Carl Lipo, who actually walked a five-ton modern version by rocking it back and forth.

An impressive—if somewhat odd—demo for sure, but many archaeologists doubt that this was a real possibility. Jared Diamond, archaeologist and writer

Gimme shelter.

Rolling Stones.

well known for his book *Collapse: How Societies Choose to Fail or Succeed*, calls it "an implausible recipe for disaster," based on the inherent instability of a stone statue whose height is five to ten times the diameter of the base.

But why did the people of Easter Island carve the heads in the first place? Late psychiatrist Anneliese Pontius of the Harvard Medical School noted that the distinctive features of the statues' bodies, in particular the emphasis of the nose, forehead, and chin, are exactly those body parts most affected by leprosy. Leprosy was relatively common on Easter Island in recent times, and Pontius suggested that the fear of the disease might have prompted carvers centuries ago to accentuate the very features damaged by leprosy—as a talisman against the disease.

Her idea never got much traction, with archaeologists and anthropologists preferring instead to view ancient Easter Island as a society where wealth, prestige, and rivalry drove competition to build more and bigger statues, not unlike the ancient Egyptians constructing larger and more impressive pyramids. We still don't know much about the significance of the statues: Why were most facing inland? Did they honor ancestors?

Did You Know . . . Most people focus on the *moai* and forget about the Ana Kai Tangata, a spectacular cave on Easter Island. Not only does the cavern offer a breathtaking view of the ocean, its walls tell us about the island's history. The red, white, and black paintings describe the tradition of the Tangata Manu, or birdman, a title awarded to whoever swam to a nearby island and collected (and brought back) the first egg that nesting birds—sooty terns, to be exact—laid that season.

Ana Kai Tangata is also popularly known as the cave of cannibals, but whether this cave was actually the site of cannibalism remains a mystery. because the translation of Ana Kai Tangata is ambiguous. Some believe it means "the cave where men eat" or the "cave that eats men."

The magnificent and strange stone heads on Easter Island represent one mystery about the people who lived there. But there's another. Despite a thriving population on the island today, in the past their civilization practically collapsed. Was it destroyed by invaders or brought on by themselves?

It's thought that statue building ended by about 1500, around the time that the sophisticated population began the two-hundred-year period of collapse. Indeed, when the first European explorers landed, in the early 1700s, Easter Island had been reduced to a population of a few thousand people. And what had been a lush forested island had become a treeless landscape—an extraordinary act of destruction, given that analysis of the traces of palm roots in the soil suggest there were once as many as 16 million trees covering the island, trees whose trunks could be 1.5 meters (5 feet) wide.

Now we can see right into his hut.

One account of the island's history claims the collapse was principally caused by denuding the forests. (It's commonly said that whoever cut down the last tree on Easter Island must have known he was doing it.) That, in turn, eliminated the supply of wood for fishing boats, fire, and logs for rolling statues—if indeed that is how they were transported—which in turn led to social unrest and conflict.

Evidence for this includes stumps and charcoal left by felled and burned trees, human bones showing evidence of violent death, and even the absence of porpoise remains. At one point, porpoises made up about a third of the islanders' diet, but by the time the explorers arrived in the early 1700s, there was no evidence of the rugged seagoing canoes that would have been necessary for a porpoise hunt.

The other scenario is quite different: invaders from Europe arrived, murdered, took slaves, and crushed the society. This is not an uncommon one—as seen in the destruction of societies in Central and South America in the 1500s. But supporters of this idea would need to show as proof that there were other causes for the deforestation of the island—that it wasn't the Easter Islanders themselves that brought about their own demise.

For instance, Hunt and Lipo, they of the walking statues, claim that rats, likely inadvertently introduced by Polynesian sailors, ate palm seeds, which prevented regrowth of the forest. There is evidence that rats gnawed on seeds found in caves on the island, but numerous other South Pacific islands have rats and palms living in peaceful coexistence. Also, two dozen other species

of tree on the island disappeared, even though they persist elsewhere in the presence of rats.

These two scenarios—the people brought the island's collapse on themselves, or that it was triggered by outsiders—are being dragged into the twenty-first century. According to Jared Diamond—and he has a number of allies on this—Easter Island is a classic case of self-inflicted environmental collapse, and we could be looking at exactly the same thing on a global scale with climate change. Countering Diamond, researcher Dale Simpson Jr., an Easter Island specialist, argues that yes, the forests were cut down, but not all at once. Farmers cleared pieces of land one by one, over decades, and rats did indeed play a role in delaying regeneration. But Europeans did the most damage by far, bringing disease and sheep, and taking prisoners. Until we understand the history of the people better, the heads are unlikely to give up their secrets.

Part 4
Curiosities and Oddities

Am I putting the toilet paper on the roll the right way?

IF YOU'RE ANXIOUS ABOUT THIS, thinking you might be putting it on wrong way around, you can relax: there really is no right or wrong. Yet there seems to be huge public interest in whether the loose end of the toilet paper hangs over the roll toward you ("over") or back over the roll away from you ("under"). In fact, Ann Landers, the advice columnist, received 15,000 letters about this topic, making it the most popular one she covered in her thirty-one-year career.

The bath and kitchen company American Standard, and Cottonelle, the makers of toilet paper, have conducted what seem to be the most authoritative surveys about toilet paper position preference, and it seems that a consistent 60 to 70 percent of those consulted prefer "over."

Explanations for this split are hard to come by. Another survey made it even more intriguing, showing that relatively few in their twenties chose "under," but that preference rose

Let's get this debate squared away.

to more than 50 percent with older respondents, then sagged again in later life. The numbers of men and women in each category stayed roughly the same.

Recently, relationship specialist and author Dr. Gilda Carle (who also has worked for Cottonelle) surveyed people and then correlated personality with preference. "Overs" tended to be a little type A, where "unders" were more relaxed. She even found that 20 percent of the 2,000 people surveyed would change the orientation of the toilet roll in someone else's house—most, of course, moving the roll from "under" to "over." And men were more irritated by paper mounted the "wrong" way.

Assuming the surveys accurately reflect public opinion, why would "over" be preferred? Is it physics? Hygiene? Or that always puzzling factor, psychology? Or, more likely, some sort of combination?

 **Did You Know . . .** A 2010 survey in England showed that toilet paper was rated the twenty-second most important invention ever, way ahead of sliced bread.

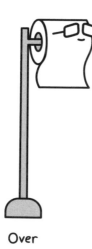

Over

Under

Undecided

There are two opportunities for physics to play a role here: first with the effort required to pull a length of paper away from the roll, and second, in the way in which a section of paper is torn. Do either of these actions vary with the direction of the roll? Let's look at the second one first.

Experience tells us that pulling straight on can lead to a wildly unravelling roll of toilet paper. Pulling to the right or left of the sheet (as opposed to straight on) will concentrate the force of the pull on the far edge. The sheet will then tear at the weakest point: the perforation. But this works the same way with both the "under" or the "over" positions.

I'm SOL.

What about pulling the paper out in the first place? In the "under" configuration it's a little more exercise for you to reach for the end of the roll, but really, as long as you're pulling the paper horizontally from the roll, usually there isn't much difference between the two configurations. Unless the roll is brand-new, that is. Physicist Jearl Walker at Cleveland State University has pointed out that when a roll is full, friction between the inner cardboard roll and the rod on which it rests is high because the roll is heavier. That makes it much more likely that a quick tug will provide . . . a single square of paper. This is where the "under" or "over" position might make a difference. If you are pulling the paper from the "under" position, it's actually supporting the roll, even lifting it a bit, allowing the roll turn more easily and lessening the chance you end up with one tiny piece. But clearly this small advantage of the "under" position doesn't have much effect on preference: the "over" camp no doubt adjusts quickly to the extra demands of a new roll.

Did You Know . . . The toilet paper company Charmin has introduced a new toilet paper roll called the "Forever Roll." And it's big. If you unrolled it from the top of a fifty-story building, there'd still be some left when it hit the ground. That's not forever, but still . . . The idea is that if you buy it—and given how thick it is, you'll likely have to purchase the special stand too—it'll last much, much longer than a standard roll. But the science! How would this huge roll influence the physics of unrolling it? Are there new physics that might be discovered? I feel a research program coming on.

So what about hygiene? Here the concern is that when you reach under the roll to get the paper, you may inadvertently touch the wall with your not-yet-clean hands. This risk is lower when the roll hangs over. In that case, at worst, you might touch some of the paper you're not yet using. But bathrooms of all kinds, public and private, are loaded with bacteria and viruses—on the toilet seats, the faucet taps, and the door handles (which are usually the most contaminated). So there's not much of an argument to made for the hygienic advantage of the "over" position. (That said, in hotels and cruise ship washrooms, the "over" position allows housekeeping to fold the first sheet into a point to signal that the room has been cleaned. That wouldn't work with the "under" position.)

If physics and hygiene concerns don't really explain the divide, it's not hard to find other anecdotal explanations: "Over" makes the end of the paper easier to grab, and it's easier to see the perforation where you want to tear it. "Under" is neater and lessens the chances of a child or a cat grabbing the roll and unwinding it all.

In fact, sociologist Edgar Alan Burns often raises the toilet paper puzzle in his classes to encourage discussion. This has given rise to even more and odder explanations for the position of toilet paper rolls: preferences were based on the breeziness of the bathroom, the distance of the paper from the toilet, or even the quality of the toilet paper. Now, those sound like rationalizations to me!

In the end, the best explanation for the passion that toilet paper position generates might be psychological. The bathroom is often the only totally private room in a house or apartment—one of the few places where you can get away from people. Some have suggested that the bathroom provides the perfect opportunity to control the environment completely, and that's when people take details like the direction of the toilet paper seriously.

But if you are still one of the undecided, consider this: the inventor of the toilet paper roll (with perforations) was Seth Wheeler. The drawings in his 1891 patent application depicted the paper in the "over" position. If that's what Seth intended, then . . .

How many universes might there be?

IF THE UNIVERSE ITSELF WASN'T HARD ENOUGH TO GRASP, we now have something dubbed the "multiverse" to grapple with. What we know about our universe is so astounding, so nearly incomprehensible, that the idea of more and different universes does seem a little over-the-top. And if you thought the word "universe" was meant to include absolutely everything, you have reason to be confused. But let's first take a step or two back.

1, 2, 3...

Part of the challenge in thinking about our universe is the distance involved. By spacecraft, it takes a few days to travel to the moon, but months to get to Mars. Mars is on average 225 million kilometers (almost 140 million miles) away. The nearest star, forty trillion two hundred billion kilometers (twenty-five trillion miles) from Earth, would take us a little over four years to reach, but we could only do that if we could travel at the speed of light—and we can't. That's a stunning distance, which, to keep things simpler, is referred to as 4.2 light-years. A single light year is 6 trillion miles, but even that distance is trivial in the context of the universe. Using all the best astronomical technologies today, we can see as far as 13 to 14 *billion* light-years away. The number of galaxies, stars, and planets that exist in that expanse is uncountable.

We're limited in what we know about our universe, not just by our current technology, but by the nature of the universe itself. Most of what we've discovered comes to us via electromagnetic radiation: this includes visible light, infrared, radio waves, ultraviolet, and microwaves. Without electromagnetic radiation, our knowledge of the universe would be a mere shred of what it is, but the usefulness of our tool is limited because light or any electromagnetic radiation has a maximum velocity.

It takes eight minutes for light from the sun to each us. That means we're seeing the sun as it was eight minutes ago. If it had blown up eight minutes ago, we wouldn't know it until now. That's just the sun. The light from very distant galaxies can take billions of years to get here, again meaning that we're seeing them as they were billions of years ago. The Hubble Space Telescope has detected light from galaxies more than 13 billion light-years away from us. That's getting close to the beginning of our universe: the best estimates are that it started as a gigantic explosion (the "big bang") 13.7 billion years ago.

This is an impressive trip back in time, but as dramatic as our current picture of the universe is, both past and present, there are two significant roadblocks to seeing further. One is that we will never be able to "see" back to the very beginning, because for the first 300,000 years or so after the big bang, the universe was too dense and hot for light to escape. We can't visually penetrate that impenetrable fog—which doesn't mean that other techniques might not get us there, but it won't be by using light.

The other difficulty is there will always be a huge part of the universe that we will likely never detect, no matter what technology we invent. The mathematical simulations of the big bang suggest that at the first moment, the split second the universe began, there was an inconceivably rapid expansion. If this truly took place, everything we are aware of, out to more than 13 billion light-years, amounts to only a tiny fraction of the total universe of which we are a part. It's as if we're a minuscule neighborhood in a vast megalopolis. The entire universe might be 100 billion trillion times bigger than all that we're aware of. One hundred billion trillion times bigger. This means that most of the universe is just too far away for light (or anything else) to have reached us in the time since the big bang. Add the even greater inconvenience that the universe is expanding—and it means those distant parts will always be beyond our reach.

So now we're speculating about what's out there. It's possible that the part of universe we're unaware of is a lot like ours, but maybe not. Maybe it is a patchwork of places where very different conditions, different kinds of physics, and even unique universes exist. There could have been other big bangs, maybe many others. They may even be happening somewhere right now, creating their own new universes.

Another of these mathematical conclusions is the "bubble" universe. In this model, universes pop up here and there, and sometimes they recede. These other universes could be extremely rare, single bubbles scattered through space, or so numerous they form a froth.

There is also the theory of a holographic universe and even a mirror universe, à la *Star Trek*—only, in this mirror universe, time itself is reversed, along with everything else. At the big bang, our universe went the way of matter; the mirror version went the way of antimatter.

Besides the purely mathematical reasons for suspecting there are other universes, here's another. It is mystifying that everything in the universe we can see seems very precisely tuned to permit life, like us. For example, at the big bang, the universe likely began to expand at a rate so fast, it can barely be calculated. Had that expansion been even a tiny fraction slower, even by one part in billions, it wouldn't have built up enough momentum, and our universe would have collapsed upon itself. Such a universe wouldn't have given rise to beings who could wonder about it, as we do. But we are here, and rather than assume that there's just one universe and it's perfectly suited to us, we could just accept that countless universes are out there, and we live in the one that actually—it's a fluke!—makes life possible.

All these ideas are speculative, but this is how the understanding of the universe moves forward. It's fun, too. Consider this: In thinking about the vastness of the "super-universe" of which our universe is a tiny part, cosmologist Brian Greene pointed out that in a volume that size, it would be very difficult to continually rearrange matter to generate unique settings. There would almost *have* to be other places that are very close to being identical to where you are right now. And a person almost identical to you, sitting and reading a book a lot like this one.

Is it true that our eyes shoot beams wherever we look?

This question may sound, well, ridiculous, but you might also be surprised at the number of people who think that when we stare at something—say, another person—our eyes emit beams that strike the person and bounce back.

This idea, known as "extramission," has a long history. It was well-known in ancient Greece, although not everyone bought into it. The mathematician Euclid apparently questioned the idea by asking, if the eyes have to shoot beams that reflect off things and return, and that's how we see, how are we able to make out the stars when we suddenly open our eyes at night? There's no time for beams to travel all the way there and back.

And today we have millions of experiments showing how sight *really* works: photons of light from around us strike the retinas of our eyes, nerve impulses are initiated, signals flow from the eye back to the visual cortex at the rear of the brain, and we "see." (In ages past, skeptics would question how the image of a mountain could shrink enough to squeeze into the eye).

Yet research indicates that belief in extramission survives. In one study, volunteers were shown various diagrams for how vision might work—light moves to the eye, light moves both ways at the same time, light bounces back and forth—and asked them which one shows how or why we see. Depending on the selection of choices, those who argued for the extramission theory varied from around 40 to 60 percent. So roughly half the people tested thought some sort of rays came from the eye, and that's what enables us to see.

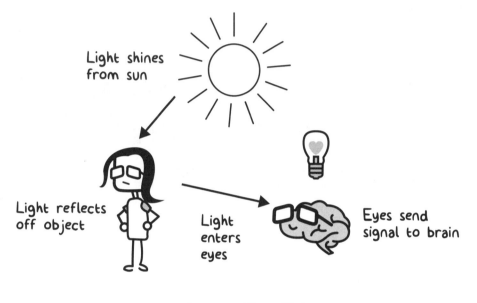

Love at First Sight

 **Did You Know . . .** "Beams from the eyes" isn't the only odd physics that people believe in. Remember when Wile E. Coyote runs off a cliff chasing the Road Runner, and then, in cartoon fashion, continues in a straight line in midair, legs churning, looks down in horror, and drops straight down? Some surveys have suggested that at least 40 percent of the population think that's exactly how it would work.

More evidence that the idea of extramission seems natural to us is the fact that children buy into the concept wholeheartedly. The belief is very common among third-grade kids but gradually dies out as they continue in school. It never completely dies, though: significant numbers of college students still think that something—rays, waves, beams, whatever—emerges from the eye when we look around us.

Science <u>Fiction</u>! *Even if we might believe in extramission subconsciously, a little scientific education should correct that, right? Well, not always. Scientist Rupert Sheldrake has made a career of challenging commonly held views in the scientific community and revising them with his own theories. Sheldrake argues for a different, modernized version of beams from the eyes: "perceptual fields, extending out beyond the brain, connecting the seeing animal with that which is seen."*

When we substitute "fields" for "beams" (there's a movie idea: *Field of Beams*), it all sounds more scientific, because there are real fields. Electromagnetic and gravitational are good examples. But there is just no convincing evidence that such fields emanate from the human brain and are somehow involved in vision.

While to some of us it seems incredible that these beliefs have persisted into the twenty-first century, recent experiments suggest that we might all, subconsciously, assume that beams do shoot from the eyes and strike the object we're looking at.

Scientists at Princeton University asked people to look at a cylinder displayed on a computer screen and decide how far it could be tipped over before falling. There was also a human face on the screen. Sometimes it was gazing at the cylinder, sometimes away; sometimes it was facing the cylinder but with eyes closed. There were different shapes and sizes of cylinders as well: some tall, some short, some thick, some thin. Many different experimental setups, but the point was to see if people estimating when a cylinder would topple over would be influenced by whether the face seemed to be looking at the cylinder. If the subjects, even subconsciously, thought that beams are emitted from the eyes, then if the cylinder was tilted toward the face on the screen, it might be able to tilt a little farther because it would be held up by those beams. If the cylinder was tilted away from the face, it might fall over sooner because of the "push" from the eyes. Think of the beam as a breeze blowing toward the cylinder.

And that's exactly what the scientists found. While not every subject allowed for pressure that "eye beams" might exert, enough did to make the results notable. The researchers were able to calculate that the force imagined by the participant would be equivalent to "a light puff of air" (for those who like numbers, a force of one-hundredth of a newton). Interestingly, none of the subjects seemed aware that they were taking this "force" into account. And only about 5 percent declared they believed in the extramission theory, so most of the participants weren't already leaning that way.

The force of her eyes is drawing me in.

This is why it's easy to believe that the eyes shoot out beams that allow us to see: it seems to be hooked into our brains at some subconscious level. And there's no doubt that this belief is so deeply entrenched—in some of us, anyway—that we likely live our daily lives automatically adjusting for the force of eye beams. A crazy thought.

Did knights have to be lifted onto their horses with cranes?

IT'S JUST A FLEETING MOMENT in the celebrated 1944 movie *Henry V*, starring Sir Laurence Olivier. Olivier, playing the king himself, mounts his horse to engage in battle. But because he's wearing a full suit of armor, he has to be lifted onto the horse with a crane! Moviegoers were likely split between feeling sorry for the horse and wondering how the king could survive if he was knocked off the horse, as weighty as he was.

But knights were never craned onto their horses. Olivier himself, both directing and acting in the movie, demanded the scene be put in, even as an armor expert tried to persuade him to abandon the idea. But the scene did raise the question of how heavy armor actually was: heavy and cumbersome enough to put a knight at risk on the battlefield?

In 2012, Graham Askew at the University of Leeds put medieval armor to the test in the lab. He recruited volunteers who were used to wearing replica armor for public performances and suited them up for tread-mill and oxygen consumption tests. The replica armor they wore were of three different styles: German, Italian, and English, all from the late 1400s, all averaging close to 35 kilos (77 pounds) in weight.

Not surprisingly, Askew determined that wearing armor made moving and breathing more challenging. In particular, armor doubled the energy cost of walking or running. In other words, to move with the same speed as they could achieve without, once suited up, knights would have to expend twice as much energy. Also, the extra weight was unlike that of an equivalent, back-pack because the legs were heavily armored as well, and swinging them was more difficult.

Askew didn't simply offer these numbers; he also speculated that the weight of armor may have been the turning point in a famous battle between the English and the French at Agincourt in 1415. French knights, heavily armored against the arrows of the English longbowmen, pressed forward across muddy, rain-soaked ground and, according to some contemporary accounts, were exhausted by the time they actually reached the English lines. They were justified in being heavily armored: the English had deployed something like 5,000 archers with the capa-bility of shooting 120,000 arrows in ten minutes. The battle was a spectacular defeat for the

French: even though they outnumbered the English by several thousand, the numbers of French killed also outnumbered the English by a factor of ten. Askew's research seemed to suggest that even as early as 1415, armor had outlived its usefulness. But armor experts were quick to dispute his findings.

For instance, the weight of the armor he used in his experiments was, according to the experts, off the charts. The average weight was 35 kilos (77 pounds), but one of them was a replica of the armor built for a seven-foot-tall behemoth named Ulrich IX, Vogt of Matsch. It's more than 50 kilos (110 pounds!). One 50-kilo suit of armor out of a total of three would definitely skew the average weight. Most armor experts peg the normal weight of suits of armor around AD 1400 at between 18 and 30 kilos (40 to 66 pounds), much less than Askew's average. Also, the armor Askew tested dates to about fifty years after the battle of Agincourt, making comparisons a little dicey.

So was he wrong? Not exactly. It's true that the French knights were heavily armored and definitely slowed further by the muddy ground they were trudging through. But on firmer ground they would likely have been fine, and strategists think that there were other reasons the French lost the battle that day.

Did You Know . . . In the early 1960s, when the United States embarked on its astronaut program, one of the pressing problems was the design of an astronaut's suit. It had to be completely sealed against the vacuum of space, tough enough to withstand the forces astronauts endured, especially while spacewalking, but flexible enough to allow freedom of movement.

One of the companies vying for the contract to build the suit contacted the British Museum, because they had heard of a suit of armor made entirely of sheet metal that might have incorporated some clever solutions to the problem of fashioning reliable, close-fitting joints. There was indeed a suit of armor that could have served as a perfect model: made for King Henry VIII, specially designed for battle on foot, not on a horse. Imagine a tight-fitting, fully flexible exoskeleton made of sheets of steel. How cool would it have been for Henry VIII's armor to be used as a model to dress astronauts for space? Sadly, the company inquiring about the armor didn't get the contract.

Besides, carrying 18 to 30 or even 35 kilos (40 to 66 or even 77 pounds) while performing physical labor is not unheard-of. Eighteenth- and nineteenth-century fur-trading voyageurs, World War I soldiers, and even modern firefighters do just that. And as far as the requirement of a crane to lift knights onto their horses, contemporary records make that seem even more ridiculous. In the early 1400s, one French knight, known as Boucicaut, apparently trained with a full suit of armor by climbing up the underside of a ladder, sprinting, jumping onto his horse, and even doing somersaults, although apparently he had to take his helmet off for the latter.

So Sir Laurence Olivier might have wanted the crane story to be true, but knights, as hefty as their armor was, definitely were able to mount their horses without cranes.

What are nanobots?

HERE'S A VIEW OF THE FUTURE: It's the year 2060. You're forty-five years old, still considered youthful in these times. You should make 150 easy. But it's not because you don't face many of the same hazards we face today, like infectious disease, cancer, and heart disease: they are all still potential threats. But they don't matter because your body is full of robotic devices heading off trouble before it starts. They're called nanobots.

These "tiny robots" are still fiction, but the idea is that they will be incredibly small. (This is not about nano-submarines staffed by humans that have been shrunk, à la the 1966 movie *Fantastic Voyage*. No shrunken people need apply in this case.) "Nano" is a billionth; a nanometer is a billionth of a meter. A sheet of paper is about 100,000 nanometers thick. The Earth compared to a toy marble is the same as a meter compared to a nanometer.

There are already hundreds of products on the market that utilize nanotechnology, from sunscreens to paint and even socks. All these have had specific atoms added to them, but nanobots are more complicated. They are robots that would be designed for tasks on the sub-submicroscopic level and assembled directly from atoms. Nanobots with medical capabilities are just one of the possibilities, but they've captured the most attention.

Their uses are only limited by our imaginations: some could patrol the circulatory system, removing any heart attack–threatening plaque buildup on artery walls; others might monitor the intestinal microbiome, eliminating rogue bacteria; and yet others could stave off the buildup of abnormal proteins in the brain that lead to Alzheimer's disease. Some could deliver drugs to specific organs or tissues, or even do nano-surgery. And how about a nanobot mouthwash? Its bots could be capable of finding food that even floss and toothbrushes can't, then dissolving a few minutes later.

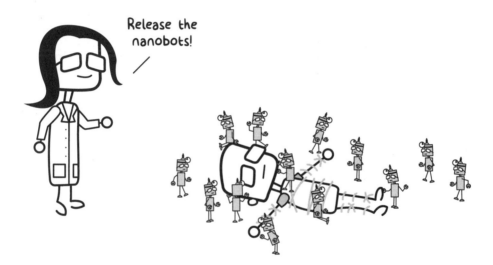

How would they be made? It's already possible to manipulate single atoms and combine them in novel ways, even to build small molecular machines. The first such machine was a molecular shuttle designed by Nobel Prize winner Sir J. Fraser Stoddart. It was, as the name suggests, able to move molecules or parts of atoms from one place to another. And that shuttle was designed and built thirty years ago! While it was nowhere near as sophisticated as future medical nano-bots would have to be, it is a reminder that these ideas have been developing for decades.

There are challenges: the smaller the nanobot, the better access it would have to any place in the body; but bigger bots would be more sophisticated and versatile—so some balance will need to be struck. Not only do they have to be powered in some way, but if that power generates heat and there are thousands of such bots in the body, dissipating that heat would be essential. And blood vessels are not placid canals: they're busy places, and having nanobots navigate to the precise target could be like paddling upstream in white water.

Where do we stand today? Not there yet but gaining ground. Here are some examples: Swiss and Israeli researchers have invented a nanoswimmer, a wire that undulates and can move its own length every second, even in the challenging environments of biological fluids (which to objects this size have the consistency of molasses). Engines will obviously be extremely important in the nanorobotic future. German scientists have built the smallest engine ever, in which a laser alternately heats and cools a trapped atom, causing it to bounce back and forth like a piston. There are other tiny engines that can generate forces larger (for their size) than muscles, and finally even nanorockets, vehicles to transport drugs around the body that propel themselves as rockets do, if not at the same velocity.

If there's a will, there's a wave.

Those are promising bits and pieces, all of which will ultimately be crucial, but so far the closest we have come to a complete robot was one developed by an international team of scientists and announced in 2018. Their nanobot used folded DNA to create a cage containing a clotting agent. The cage protected the agent, thrombin, from attack by the immune system, and delivered it to tumors. It caused the blood vessels supplying the tumors to clot and shut down, killing the cancer cells.

This is a pretty dramatic demonstration of the potential of medical nanobots, even though the safety and effectiveness will have to continue to be tested on other animals before being tried in humans. But we're certainly closer than we've ever been to realizing the potential of the nanobot to keep us healthy and long-lived!

Do aliens exist?

WHAT COULD BE MORE FASCINATING, and even a little scary, than the idea that there are "others" out there, alien civilizations much, much more technologically advanced than ours? But if they are out there, how can we find them?

Shhhh...

There have been really two ways of trying to prove their existence: one is to argue that mysterious structures on Earth like the Great Pyramid of Giza or the Nazca Lines were built by aliens who visited here thousands of years ago. These claims have sold a lot of books but have been proven wrong—every one of them. The other is the scientific approach: the search for evidence of intelligent civilizations in the vastness of the galaxy. Part of that search has begun by locating planets around other stars; part has been searching for radio waves that might be signals from aliens.

So far nothing has turned up. But we have been looking for only about fifty years, and the universe is incomprehensibly vast. One off-beat idea is that we haven't been looking in the right place. What if alien civilizations had set up shop much closer, even in our own solar system? Or right here on Earth? What if other civilizations visited Earth, not thousands of years ago, but hundreds of millions of years ago and settled here? Would we be able to find any trace of them today?

That's a tough question. There wouldn't be much left by now: most of the traces would have been long since ground into dust, eroded into oblivion, or buried under shifting tectonic plates. The search would have to be much more sophisticated than just digging and hoping.

However, there is something we could do by using ourselves as a reference. Seven billion humans are changing the Earth today. In the last two hundred years, our industrial activity has caused a sudden increase in the amount of carbon dioxide in the atmosphere; dramatic changes in the way nitrogen cycles through the atmosphere, soil, and rivers; and the appearance of plastics everywhere, especially in the world's oceans and in the living things there. Speaking of oceans, sediments laid down in them are infused with metals released from industrial processes and mining. It begs the question: Could alien civilizations have made changes like these and left a mark that we could find?

It would be tricky. On one hand, if that long-ago civilization destroyed itself in war or persisted in destructive industrial practices, they might have left us traces of their existence. On the other hand, their non-natural materials, like our plastic, might have decomposed by now. And even if

They left without a trace!

we found proof of chemical changes in the atmosphere millions of years ago, these might have had a natural origin rather than being the result of a past civilization. Suffice to say, it would be hard to come up with any persuasive evidence that aliens resided on Earth. But what about our solar system neighborhood?

The moon, Mars, or the asteroids might be the easiest places to search, but doing that without actually being there is challenging: we know very little about them. For instance, even though the Lunar Reconnaissance Orbiter has surveyed close to 100 percent of the surface of the moon with unprecedented resolution, a 5-meter-wide (16 foot) lump of rock wouldn't look much different than a 5-meter-wide satellite of alien origin. And while we're somewhat familiar with the moon, the asteroid belt is full of objects we really know nothing about.

Bedrock   Shamrock   Alien rock

Even landing on any of these familiar parts of the universe and searching for signs of aliens would still be no guarantee of success. On Mars, shifting sands and strong winds would, over time, bury any evidence, and on the moon and the asteroids, constant bombardment by meteorites might pulverize and/or bury any trace of alien beings there.

Of course, the easiest way to discover aliens is if a piece of alien technology visited us now—and I'm not talking about UFOs! The sensational discovery of what's called 'Oumuamua might be an example.

'Oumuamua means "visitor from the distant past" in Hawaiian. It was first discovered by Pan-STARRS (Panoramic Survey Telescope and Rapid Response System) in Hawaii. A high-velocity object first seen in 2017, it entered the solar system, passed by the sun, and now is on its way out of the solar system again. Astronomers realized 'Oumuamua was peculiar: it was a thin, pancake-shaped object, neither comet nor asteroid, swinging by the sun at speeds that couldn't be accounted for by the sun's gravitational pull. It's the first object ever seen in our solar system that came from somewhere else in the universe.

Dr. Abraham "Avi" Loeb, an astronomer at Harvard University, argued that this space pancake might actually be a solar sail, a very thin structure maybe no more than a millimeter thick (more of a space crêpe) that would gradually accelerate through space pushed by radiation from the sun. Solar sails don't occur naturally. Scientists on Earth have been designing them for space exploration, and if this was one, it would have been manufactured by an alien civilization.

Dr. Loeb went on to speculate that 'Oumuamua might be a defunct piece of machinery traveling through space, like a satellite that's run out of power but is still orbiting Earth. It sent no message. Or it could just be some other weird, but natural, space object (which most experts think).

Unfortunately, we'll never know: 'Oumuamua will likely never return. But we will still keep searching for aliens, near or far away.

How far underground can living things be found?

When Alice followed the white rabbit down the rabbit hole to Wonderland, she fell straight down for a very long time but eventually at the bottom discovered a spectacular variety of people and creatures. While Lewis Carroll populated his underground world with creatures of his imagination, there are actually plenty of living things that are thriving in the soil beneath our feet and deep below the surface of our oceans. But how far down does life go?

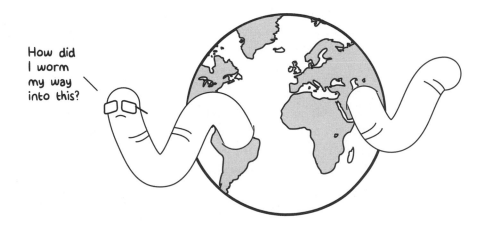

How did I worm my way into this?

The deeper down you go, the less familiar the creatures, so let's at least start with some we know. One giant species of earthworm burrows down about 3 meters (10 feet), but the deepest burrows on Earth belong to Nile crocodiles: they can descend close to 12 meters (almost 40 feet). And that's pretty much the farthest any creature *digs*.

Of course it's not always necessary to dig to penetrate far below the surface of the Earth: the extreme depths of the ocean are about 11,000 meters (almost 7 miles) down, and the Mariana snailfish holds the depth record for all fish, able to live at least 8,400 meters below the surface. It isn't much to look at, a few inches long and semitransparent, but the Mariana snailfish withstands pressures at those depths that have been described as equivalent to an elephant standing on your thumb—about a thousand times the pressure exerted by the Earth's atmosphere.

Did You Know . . . The naked mole rats of East Africa don't burrow very deep—only about half a meter below the surface—but they make up for depth with incredible breadth. These creatures, bigger than a mouse but smaller than a rat, dig vast networks of underground tunnels that can stretch thousands of square meters!

There have never been any fish found below 8,400 meters in the ocean, even though that depth is only about 75 percent of the way to the deepest ocean floor. Yet it might not be those crushing pressures that are the reason but rather the body mechanisms fish need to maintain the correct internal balance of salt and water at those pressures.

While fish are in a holding pattern at about 8,400 meters, other organisms are not. A few years ago a robot vessel descended to the bottom of the Mariana Trench, that deepest part of the ocean 11 kilometers down and recovered both copious populations of bacteria (populations much denser even than in comparatively shallow waters) and even tiny shrimp-like amphipods—life at the very bottom of the ocean.

Of course, deep diving in the ocean, pressure challenges notwithstanding, is easier than drilling down into the earth—or, worse, finding places to live in the rocks. But there are creatures that manage to do that. While South Africa is not the only place to find these rock dwellers, its deep mine shafts are where most have been discovered.

Most are simple animals, like roundworms. Several species have been found between 2 and 3 kilometers (nearly 2 miles) down, where it was thought that high temperatures, great pressures, and lack of oxygen and food would make multicellular life impossible. Yet water has probably soaked through the rock in these life-bearing areas, and one roundworm, or at least pieces of its DNA, was found at 3,600 meters below the Earth's surface, a place where the temperature would be 48° Celsius (118° Fahrenheit). Apparently these roundworms and other animals found at equivalent depths in South Africa are, like the amphipods in the Mariana Trench, grazing on sheets of bacteria.

And bacteria must be acknowledged here. Even though they're microscopic single cells with few interesting behaviors, they represent an absolutely vast underground population beneath the land *and* the sea. In other words, they occupy the entire Earth's crust. For instance, a survey of microbes after eruptions of an undersea volcano called Axial Seamount, 500 kilometers (more than 300 miles) west of Oregon, revealed whole new groups of bacteria, never before seen, that had been blasted out of Earth's crust into the ocean 1,500 meters (more than 4,900 feet) below the surface of the water. Wherever animals or fish are found, bacteria go deeper. And there's every reason to think they have existed out of sight, underground, for thousands if not millions of years, perhaps going back to a time in Earth's history when these underground microbes accounted for ten times as much life on Earth as those living on the surface.

We've had evidence for only the last thirty years that the Earth's crust is a giant incubator, and despite these new and surprising discoveries, some scientists think that high temperatures and pressures would make 10 kilometers (roughly 6 miles) the maximum depth to support life. But maybe that's just traditional thinking. If there's water, there might be life.

And the question of how much deeper life can go leads to yet another intriguing question: If life can have existed below the surface of the Earth, out of sight of the sun, out of touch with oxygen, for hundreds of millions of years, why couldn't the same thing have happened on other planets, especially Mars? Mars looks inhospitable now, but it was much more Earth-like in the past. If life originated then, it might still be there under the sands of that planet. The thought of drilling kilometers down into the Martian crust seems far-fetched at the moment, but there aren't many scientists who would bet against that kind of life on the Red Planet.

Could we be cryonically frozen and then revived centuries later?

WALT DISNEY DIED IN 1966 and was cremated two days later. But ever since, there have been persistent rumors that he had actually been cryonically preserved—that is, his body was being kept frozen until some possible future when he could be warmed up and come alive again.

The reason the myth of Disney's cryopreservation, as it's called, has lived on despite the total lack of evidence must be because it fascinates us. Imagine being frozen the moment you die and

I'm a bit chilled.

then brought back in the year 2120, or 2300, or whenever. It would be time travel! But obviously it's never been done, and it raises serious questions, like how and why.

First the how. It seems encouraging that there are many animals that can be dramatically cooled down but still survive. For instance, when Arctic ground squirrels hibernate, their body temperature drops dramatically, sometimes to below the freezing point of water. Parts of their brains shut down, and they remain in that near-death state for months. But when spring comes, they are back to life in every way. Even more dramatic are wood frogs, which actually freeze solid in their winter. They, too, are fine come spring.

So there is some evidence that freezing, then thawing, can work, although these are not totally convincing examples, because (a) we're not ground squirrels or frogs, (b) they're only super-cooled or frozen for a few months, not centuries, and (c) they obviously have special metabolic tricks that make this possible.

Yet there are a couple of intriguing medical cases that suggest it might not be impossible for humans. In 1986 a two-and-a-half-year-old girl fell into a creek and wasn't rescued for sixty-six minutes. She had been submerged all that time, and when she reached the hospital her temperature was 22° Celsius (72° Fahrenheit), 15° Celsius (27° F) below normal body temperature. She wasn't breathing and her heart wasn't beating, but after a series of intense medical interventions she recovered, and other than a slight tremor in her hands, she is pretty much normal today. There have been other cases, including a child and man whose body temperatures were in the 20s Celsius (high 60s F) when discovered. All have recovered.

But 22° Celsius (72° F) isn't minus 196° Celsius (minus 321° F), the temperature of liquid nitrogen in which cryonically preserved bodies would need to be kept so that chemical reaction rates in the body are so slow—9 octillion times slower than normal—that they essentially don't happen at all. (Those wood frogs? If they were kept at minus 196° Celsius, they'd shatter if they hit the floor.)

 DON'T TRY THIS AT HOME! Early Canadian explorers found frozen frogs in the ground when they dug latrines. Weirdly, they carved initials in them, and when spring came, the frogs revived but still bore the imprints of the initials. Also, it's possible to drop a wood frog into a small container of water, put it in the freezer, and as long as the temperature remains above minus 5° Celsius (23° F), the frog will be fine when the cube is thawed. Still, probably best not to try.

Most animals can't survive freezing because the uncontrolled formation of ice crystals in their tissues rips these apart, resulting in death. Proponents of cryonics argue that "vitrification," a process in which the viscosity or thickness of fluids in the body is increased, can be used to prevent this crystallization. Substances called cryoprotectants, similar to antifreeze, are required for vitrification, but that's not so strange, because those animals that can survive extended periods of cold temperatures often rely on their own natural cryoprotectants.

The scenario is something like this: if you opt for cryopreservation, upon "dying" your body is cooled to about 0° Celsius (32° F), cryoprotectants are added, and cooling continues until

I'm starting to feel liquidy.

I'm feeling quite solid.

about minus 120° Celsius (minus 184° F). At that point, your body liquids and structures are so viscous, somewhere between liquid and solid, that they won't be damaged as you are cooled to minus 196° Celsius (minus 205° F). And there you lie inert, dead in appearance but with the potential of revival.

While it's possible your entire body could be preserved this way, we don't know yet if, upon revival, it would be completely intact. The simpler the organ, the likelier it could survive. Already scientists have been able to freeze mouse ovaries to minus 196° Celsius (minus 321° F), thaw them, and transplant them into female mice to produce live, normal babies.

Also, slices of brain tissue from rats that have been vitrified to minus 130° Celsius (minus 202° F) and warmed up seem, at least with limited testing, to have revived and functioned. But what about the whole brain? If you wanted to be "reawakened" in the year 2200, wouldn't you want your brain, which contains all your thoughts, feelings, memories—the entire *you*—to be the same as it was before you went under? The brain has 86 billion neurons, or brain cells, and an equal number of other kinds of cells. Those neurons make tens of thousands of connections with each other. Lose any significant number of cells or connections and, at the least, you'd have serious amnesia. At worst, you'd have returned in a permanent coma. And even if the neurology of the future would be able to reconstruct an intact brain from the remnants of yours, the product wouldn't be you but another, new human being. What is the point, then?

Futurists seem confident that the medicine of the future will allow the 3-D printing of replacement organs and ultramicroscopic robots to circulate through veins and arteries and fix things as they break down. If all this happens, there could come a time when we might be able to survive freezing more or less intact.

 **Did You Know . . .** Walt Disney didn't do it, but so far one of the main companies in the cryopreservation organization in the United States, Alcor Life Extension Foundation, advertise that they have recruited 1,246 members (people who have taken all the steps necessary to be cryopreserved but are still alive) and 170 "patients" who are already being preserved.

But even if we are able to awaken as our old selves, there will still be challenges to this second life. What if the world to which we return is so foreign and unrecognizable that it is just too difficult to adjust? A world with no friends, nothing familiar, and nothing to do—or, worse, a world that seems much less livable than the one you left. We might find that just because the technology allowed us to do it, cryonic rebirth is not what we were hoping for.

History Mystery

When was the wheel invented?

While we may not know exactly who invented the wheel, we can estimate when the wheel was invented, although this is a much more difficult question to answer than you might think. Obviously the wheel has been around for a very long time, several thousand years at least. It's also apparent that today we are a totally wheel-centric civilization. Unless you're out alone in the deep woods, you will usually see some sort of wheel within a few seconds of looking. But it wasn't that way long ago, and several lines of evidence have to be pieced together to pinpoint when the wheel first appeared in human history.

5,500 years ago

Those include language, archaeology, and horses. Each alone is not enough; together they tell a story of how the wheel's influence moved from agriculture to war.

First, language. The word "wheel" in English can be traced back to a word that meant "the thing that turns" in an extinct language called Proto-Indo-European. This language was spoken between 5,500 and 4,500 years ago and then died out (but not before giving rise to "daughter" languages, like English). There are certainly no words for wheel before 6,000 years ago, and for this reason, the best guess is that the wheel was invented sometime around 5,500 years ago, circa 3,000 BCE. There are other, related words in Proto-Indo-European like "axle," "thill" (the pole connecting the harness of the animal pulling the wheeled vehicle), and a word roughly meaning "transport by vehicle." So, by this time, some of the crucial challenges in building a wheeled vehicle had seemingly been solved.

The language was widespread, so we can't pinpoint where the first wheels were being used, but that's where archaeology comes in. It has unearthed drawings or engravings of wheeled vehicles and actual remnants of wheels themselves. The first wheels were solid pieces of wood: usually not cross-sectional slices of wood from the trunk of a tree but vertical slices cut into a circular shape.

If you'd asked that farmer for directions, we'd be at the market by now!

While there may have been challenges shaping that wheel, the real challenge was the mechanics of the axle. The first wagons had immobile axles and the wheel turned around them. The fit had to be perfect: if the axle was too tight, the friction would slow down the wagon dramatically, but if too loose, the wheel would wobble. That required precision carpentry skills and tools.

The finished wagon, with poles attached to a yoke for the animal (probably an ox) dragging it, was pretty heavy. Apparently these first users of the wheel tried to reduce weight by making smaller, lighter axles to create less drag. Because these were more likely to break, they reduced the weight of the cart by making it narrower. Some were only a meter (3 feet) across. Turning a wagon into a cart by using two wheels rather than four reduced drag even more. Eventually a rotating axle fixed to a wheel replaced the fixed axle, and wagons and carts could get bigger again.

Did You Know . . . There is no evidence that Central American civilizations like the Aztecs or the Maya ever used wheeled vehicles. Some have suggested that was because there were no large draft animals suitable for pulling carts or wagons. What's curious is that they definitely knew about wheels but they only used them for children's toys.

It's hard to imagine the impact the wheel had on life back then: wagons and carts revolutionized farming. Now heavy loads could be carried long distances, allowing communities to spread out over the landscape.

Indeed, we find all kinds of evidence of the wheel starting roughly a hundred years after it was first invented: pictures of wagons and carts, models of them, and even physical remains of wheels, axles, and pieces of carts. These artifacts appear in a variety of locations, from the Middle East to northern Europe. A perfect example is an accurately dated image of a wagon on a cup found in southern Poland. It's somewhere between 5,500 and 5,350 years old, making it the oldest such image anywhere.

Two other dramatic developments changed the course of the wheel forever. One was the domestication of the horse. Skeletal remains of horses from around 3000 BC show evidence of wear on the teeth—wear that would have been caused by the bit a rider would put in a horse's mouth to control it. Their impact, like that of the wheel, was enormous. It's been estimated that a shepherd with a dog can herd about two hundred sheep; with a horse and a dog, five hundred. But horses weren't just ridden by shepherds; they were harnessed to chariots.

The invention of the chariots about 4,000 years ago was made possible by a significant change in wheel design: spokes. Spoked wheels were much lighter than solid wooden ones, which meant that vehicles could be built for speed. The chariot itself was stripped of any unnecessary weight. The two-wheeled chariot drawn by one or two horses, with a single driver armed with javelins, changed warfare as soon as it was introduced.

From that time on, the wheel has been an essential feature of civilization. Today, 5,500 years after they were first invented, wheels can be found everywhere on Earth—and even on the moon and Mars.

Science _Fact!_ There used to be an old phrase popular among biologists: "Nature never invented the wheel." At first it makes total sense. The first wheels spun freely on an axle, unconnected except for pins to prevent them from spinning off. But every part of a living organism must be connected to the body to be supplied by blood, powered by muscles, and connected to nerves. How could a spinning organ do that without tying everything in knots?

But these arguments are irrelevant now, because there _is_ a wheel in nature. Countless species of bacteria swim by spinning a rigid, coiled flagellum. It doesn't wave back and forth, it turns, propelled by a tiny living motor. It's the only living wheel ever discovered.

I'm au naturel

Acknowledgments

This is the fourth book in a series that was never intended to be a series, and there are many people who have stuck with it throughout. They were instrumental in making the first three possible and have done the same for this one.

On the Simon & Schuster side, Kevin Hanson, Sarah St. Pierre, Catherine Whiteside, and Nita Pronovost were once again both creative and supportive. And of course the entire staff of Simon & Schuster deserve credit for coming up with a continuing flow of intriguing questions.

I think it's fair to say that my editor Meg Masters and I are of one mind when it comes to the ingredients of an interesting story—it's just that she can say it in fewer and often better words.

My agent Jackie Kaiser and also Stephanie Thompson and Meg Wheeler at Westwood Creative Artists have worked hard to ensure that these books exist at all and that they reach an audience outside Canada.

Special thanks to Lucia Jacobs, Mark Changizi, Graham Askew, Corey Keeble, Toby Capwell, Dale Simpson Jr., and Ken Storey, all of whom contributed essential inside information on a variety of topics.

On the research side I especially thank University of Guelph physicist Joanne O'Meara and wildlife biologist/writer Niki Wilson. They both seem to be resigned to the fact that I will ask them for help out of the blue and are extremely accomplished at delivering exactly that.

And where would a writer be without friends and family chiming in? I've named them before but they deserve recognition again: Rachel, Paul, Max, Kait, Amelia, and Brendan; JJ, Finn, and Grace; the Flathead group, The Beakerband, and their partners; and The Men of the Road and their partners.

And in more ways that I can count, none of this would have happened without Mary Anne.

Jay Ingram has written seventeen books, including the bestselling three books in this series, *The Science of Why*, *The Science of Why²*, and *The Science of Why, Volume 3*. He was the host of Discovery Channel Canada's *Daily Planet* from the first episode until June 2011. Before joining Discover, Ingram hosted CBC Radio's national science show, *Quirks & Quarks*. He has received the Sandford Fleming Medal & Citation from the Royal Canadian Institute, the Royal Society of Canada's McNeil Medal for the Public Awareness of Science, and the Michael Smith Award for Science Promotion from the Natural Sciences and Engineering Research Council of Canada. He is a distinguished alumnus of the University of Alberta, has received six honorary doctorates, and is a Member of the Order of Canada. Visit Jay at **JayIngram.ca.**

🐦 @jayingram

Itching for more science facts and fictions?

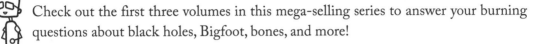 Check out the first three volumes in this mega-selling series to answer your burning questions about black holes, Bigfoot, bones, and more!

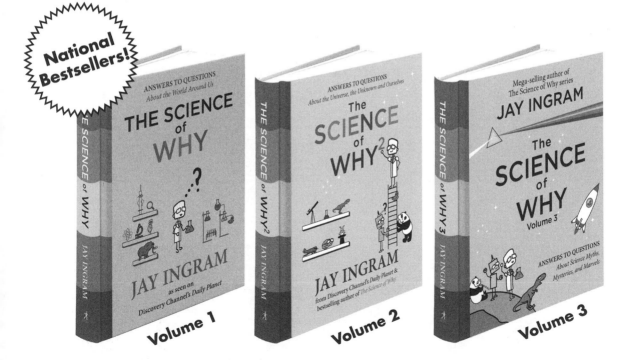

National Bestsellers!

Volume 1

Volume 2

Volume 3